
MOULD & CORE MATERIAL FOR THE STEEL FOUNDRY

MOULD & CORE MATERIAL
FOR THE
STEEL FOUNDRY

by

A. D. SARKAR, M.Eng., B.Sc.

PERGAMON PRESS

OXFORD · LONDON · EDINBURGH · NEW YORK
TORONTO · SYDNEY · PARIS · BRAUNSCHWEIG

Pergamon Press Ltd., Headington Hill Hall, Oxford
4 & 5 Fitzroy Square, London W.1
Pergamon Press (Scotland) Ltd., 2 & 3 Teviot Place, Edinburgh 1
Pergamon Press Inc., 44–01 21st Street, Long Island City, New York 11101
Pergamon of Canada, Ltd., 6 Adelaide Street East, Toronto, Ontario
Pergamon Press (Aust.) Pty. Ltd., Rushcutters Bay,
Sydney, New South Wales
Pergamon Press S.A.R.L., 24 rue des Écoles, Paris 5^e
Vieweg & Sohn GmbH, Burgplatz 1, Braunschweig

First edition 1967
Library of Congress Catalog Card No. 67–24316

Printed in Great Britain by A. Wheaton & Co., Exeter

(3347/67)

CONTENTS

PREFACE

A LOOK at the various types of mould and core material available today will show that there has been a steady progress in the development of that important aspect of steel founding. About a decade ago when the carbon dioxide process was finding application in this country, zealous foundrymen described it as a new epoch. This was later supplemented by the air-set materials and there is a continuing interest in this field of chemically hardened moulds and cores. I myself, among others, have often wondered if clay-bonded sands would soon become a thing of the past. However, the trend appears to be that the natural sands and the synthetic greensand mixes are going to stay for a long time unless the non-clay-bonded materials become cheaper and more plentiful than they are today.

The first conditions which a mould or core material must fulfil are that it must be workable, i.e. it must have a degree of flowability and that it must have adequate strengths for the constructed moulds to retain their rigidity. Much work has therefore been done to understand the filler grains and the bonding media in the green state. It is equally important that the mould–core surfaces are able to resist the washing action of the flowing metal during casting and remain rigid until the casting has solidified or to breakdown at an appropriate stage when the casting is contracting.

While writing this book I have assumed that a reader unfamiliar with mould and core making principles will read the first chapter of my book on the carbon dioxide process, published by Pergamon Press. In this book the subject of mould and core material is considered in terms of their green properties and the chapter on sand testing summarizes the more established parameters used in most steel foundries for assessment of foundry sands. One theory of the mechanism of green bonding is given and the pertinent aspects of processing foundry mould and core materials are also discussed. Of the casting defects, scab and hot tear have received much attention, and two individual chapters are given to them. The high temperature properties of mould and core materials have always been a favourite subject with me. There is a very large number of papers given by various workers on the subject of hot strengths of foundry sands, but it is doubtful if a correlation between these properties and casting defects has been definitely established. Sand technicians are anxious to find suitable acceptance tests on high temperature properties which would give some definite indication with regard to the behaviour of hot mould and core interfaces in terms of casting defects. This subject should be pursued in close collaboration with production castings. I have discussed the aspects of hot and retained strengths in my book on the carbon dioxide process, and the possible role of high temperature properties on cast quality has again been indicated in this work wherever necessary together with a chapter giving a brief résumé of hot strength studies by research workers.

For the last 4 years I have been in the employment of Head Wrightson and Company Limited, and I was given the opportunity of gaining first-hand information on castings made in plain carbon and alloy steels employing the most up to date mould and core making techniques in their steel foundries. This gave me the background to attempt a book of this kind, and I am grateful to them. I acknowledge my thanks to British Industrial Sand Limited, Ridsdale and Company Limited, Pneulec Limited, August's Limited, F. & M. Supplies Limited, Harborough Construction Company Limited, the Fordath Engineering Company Limited, Catalin Limited and Sternol Limited for giving permission to reproduce illustrations and information. Table 25 is reproduced from B.S. 2564:1955 by permission of the British Standards Institution, 2 Park Street, London, W. 1. My thanks are due to the United Steel Company Limited for permission to reproduce Fig. 15 and the relevant curves in Fig. 42, to Professor Richardson for Fig. 8 and to Mr. J. M. Middleton for Fig. 41. I thank Mr. F. LeServe, Mr. R. Chadwick and Mr. Ridsdale for showing an active interest in this work. Mr. C. Wough kindly made the rough sketches for Figs. 22 and 25.

As with the manuscript of my previous book, Pergamon Press carried out the final drawings from the very rough freehand sketches of the diagrams I submitted. I am greatly indebted to them for giving me this invaluable assistance.

A. D. S.

CHAPTER 1

INTRODUCTION

THE basic function of a mould is to provide a cavity which would receive liquid metal to produce a casting of a desired shape and size when the metal cools by solidification. A core is a solid object which is inserted in the mould cavity so that metal flows around it and, upon solidification, the casting is devoid of metal in the part occupied by the core initially in the mould cavity.

Moulds can be permanent moulds such as those used in die-casting and expendable ones such as those produced from refractory aggregates. The permanent moulds can be used many times whereas the latter expendable moulds must be constructed separately each time a casting is to be produced. Permanent moulds are usually metal moulds and possibly the only permanent mould material used to receive steel is grey or nodular cast iron such as ingot moulds. However, steel ingots produced in steelworks from ingot moulds do not, by usage, come under the category of castings, and a steel foundry is essentially that branch of engineering where shaped components are produced by using moulds and/or cores. The basis material for these are some refractory particles such as silica sand.

The various forms of mould and core making methods have been described elsewhere,* and the main requirements of a mould or core material are that:

(1) It must be amenable to shaping processes such as by ramming around patterns and in core boxes;

* A. D. Sarkar, *Foundry Core and Mould Making by the Carbon Dioxide Process*, Pergamon Press, Oxford, 1964.

(2) The rammed structures must be able to retain their shapes and rigidity once constructed until the metal cast in the mould has formed a strong solidified skin.

That is, the material must flow readily for compaction and the refractory particles such as silica sand grains must be held together by some bonding medium. Consider the sand grains as spherical in shape like billiard balls. When two such particles are brought together they simply touch at the points of contact, but little or no force is required to separate the grains (Fig. 1a). However, if a bonding medium is shared by the two grains near the point of contact (Fig. 1b), an appreciable force would be necessary to separate the grains.

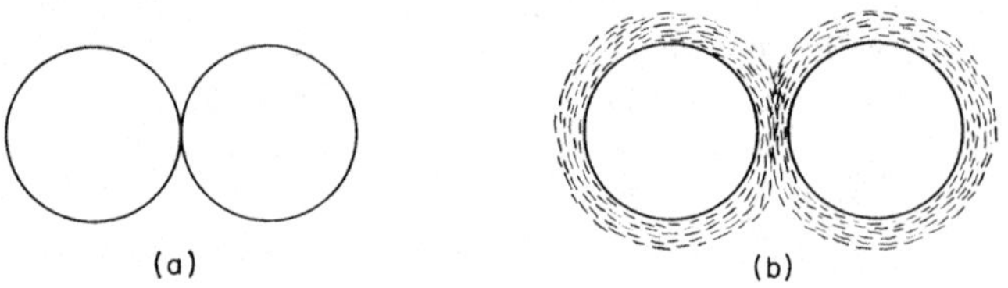

FIG. 1. Two sand grains.
(a) Without bond. (b) With bond.

The bonding medium in a moulding material is distributed around the grains so that, as far as possible, the particulate matters are completely coated by the bonding agent. This is necessary because, since a mould involves the use of many sand grains, no matter how the particulate matters are oriented, there will always be an amount of bond available at all points of contact ensuring uniform strength at all parts in the compacted mould. The distribution of the bonding agent around the sand grains is effected by mixing or milling the refractory particles with the bonding medium in sand mills.

The various types of bonds employed in the steel foundry are:

(1) Clay bond.
(2) Oil and organic bond.
(3) Thermosetting resin bond.

(4) Air-setting bond.
(5) Sodium silicate bond.
(6) Ethyl-silicate bond.

The most widely used basis material in formulating mould and core materials for the steel foundry is silica sand together with some non-siliceous materials such as zircon and chromite.

The basis refractory aggregates have maximum flowability which diminishes as the silica sand is mixed with the bonding agent. However, the moulds as prepared must have adequate strengths, i.e. a definite amount of bonding agent must be present. The amount of bond is optimized so that the moulds have adequate strengths without a drastic loss of flowability.

The mould material is prepared in a sand mill which is only a part of the whole sand plant system described in Chapter 7. To obtain satisfactory castings, it is necessary that each batch of mould material is tested to ensure consistent properties. The probable strengths of moulds are expressed by measuring the compressive strengths of samples in the laboratory. A measure of flowability is by knowing the shatter index value which allows a compacted specimen to fall through a height. If the specimen breaks into small individual grains the mould material has maximum flowability with no bond and zero shatter index. Since there must be an optimum bond and a good flowability, a certain value of the shatter index is regarded as the optimum. Moulds are also used in the dry state for higher strength than that obtained from green moulds and dry compressive strengths are measured. Certain core materials such as oil- and silicate-bonded sands have very high strengths, and tensile strengths are used in these cases for property control tests. A further parameter measured is the transverse strength. As liquid steel enters the mould, gases are evolved. The sources for the gaseous matters may be steam when the mould is green, i.e. not dried or the volatilization of carbonaceous matters such as in oil- or organic-bonded core materials. These gases must be extracted from the mould–core interfaces through the body of the compacted materials, otherwise the gases will be trapped in the casting giving gross defects. Although venting channels are provided in cores while constructed, there should be voids between the adjacent sand grains to allow the gases to permeate out of the cores into the atmosphere. The amount of voids present is measured by noting the permeability of the mould material.

The properties such as green and dry strengths, permeability, amount of gas evolved, shatter index together with hardness of rammed moulds, are measured on a routine basis, and the testing procedures are outlined in Chapter 2. However, the properties of a mould when heated are also important, and parameters such as hot compressive strengths, i.e. the compressive strengths of specimens while heated to a temperature, should also be measured.

CHAPTER 2

SAND TESTING

THE terminology sand testing is loosely used by foundrymen generally meaning testing of mould and core materials as prepared in the sand plants, but also involving the assessment of certain physical characteristics of, for example, incoming silica or naturally bonded* sands such as grain size. There is a continuing effort to devise testing methods in an attempt to understand the behaviour of mould and core materials at room and elevated temperatures. For example, British Cast Iron Research Association recommend a simple impact test on Izod lines, and claim better results than those obtained from testing for shatter index. In this chapter only the testing procedures of the more commonly used properties such as green and dry compressive strength, shatter index, green and dry permeability, clay content, sieve distribution, etc., are summarized. Certain high temperature properties such as hot compressive and retained strengths are not generally measured on a routine basis, and apparatus used by different foundries may vary. The method used by the present author for evaluating hot compressive strengths is outlined in Chapter 15.

The testing procedures described below are designed for both routine control and research purposes and are in accordance with the specifications laid down by the American Foundrymen's Society (AFS).

* Silica sand is basically the mineral quartz and it is necessary to add bond such as clay to hold the grains together. Naturally bonded sands occur in nature where the silica grains are already mixed with a clay and hence can be used directly as a moulding material without adding any bond.

Green Compressive Strength

This may be defined as the maximum load required to break a green mould under axial compressive stress. The load is expressed in pounds per square inch and is applied on a specimen 2 in. diameter × 2 in. high having a standard ramming density.

The specimen is prepared by using a standard rammer and a specimen tube (Fig. 2). The AFS standard sample is prepared by ramming a known weight of sand with three blows by making a 14 lb weight fall a distance of 2 in., the exact weight of sand being found by trial and error. For example, a silica sand bonded with bentonite* may need 165 g, while certain naturally bonded sands may require as much as 195 g to prepare a standard specimen.

The specimens are ejected from the tube by using a stripping post.

For mould materials of low green strength (up to 1 lb/in^2) or specimens of high ramming density, ejecting a specimen by the stripping post is not successful and a split specimen tube is used. The compressive strengths may be measured on a deadweight type Universal sand testing machine (Fig. 3).

The Universal Strength Machine

The machine can be fitted with various accessories to test dry compressive, shear, tensile and transverse strength. For all dry strength measurements it is usual to dry samples for 2 hr at 110°C. If a higher temperature than this is used this must be reported in the sand-test results.

The compressive specimen is crushed between two compression heads, which are designed to be self-aligning. The transverse specimen is 1 in. × 1 in. × 6 in. long and is broken between two knife edges. The tensile specimen has a cross-section of 1 in^2 at the centre (Fig. 4). The toughness of a green sand is often measured in terms of total deformation multiplied by the green compressive

* Bentonite is a clay.

(*Courtesy Ridsdale and Company Ltd.*)

FIG. 2. Standard rammer and specimen tube.

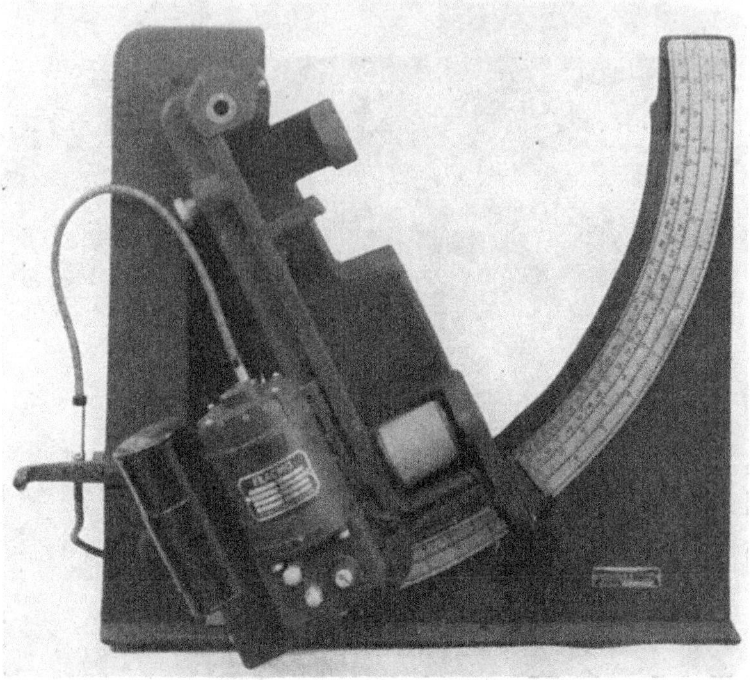

(*Courtesy Ridsdale and Company Ltd.*)

FIG. 3. Universal strength machine.

strength, and gives a good indication of the workability of the sand.

Green Permeability

Permeability is that property of a mould or core which allows air or gas to escape through its pore spaces, and is measured by noting the rate of flow of air under a known pressure through a standard AFS compressive specimen. The permeability number p is defined as follows:

$$p = \frac{vh}{pat}, \quad (1)$$

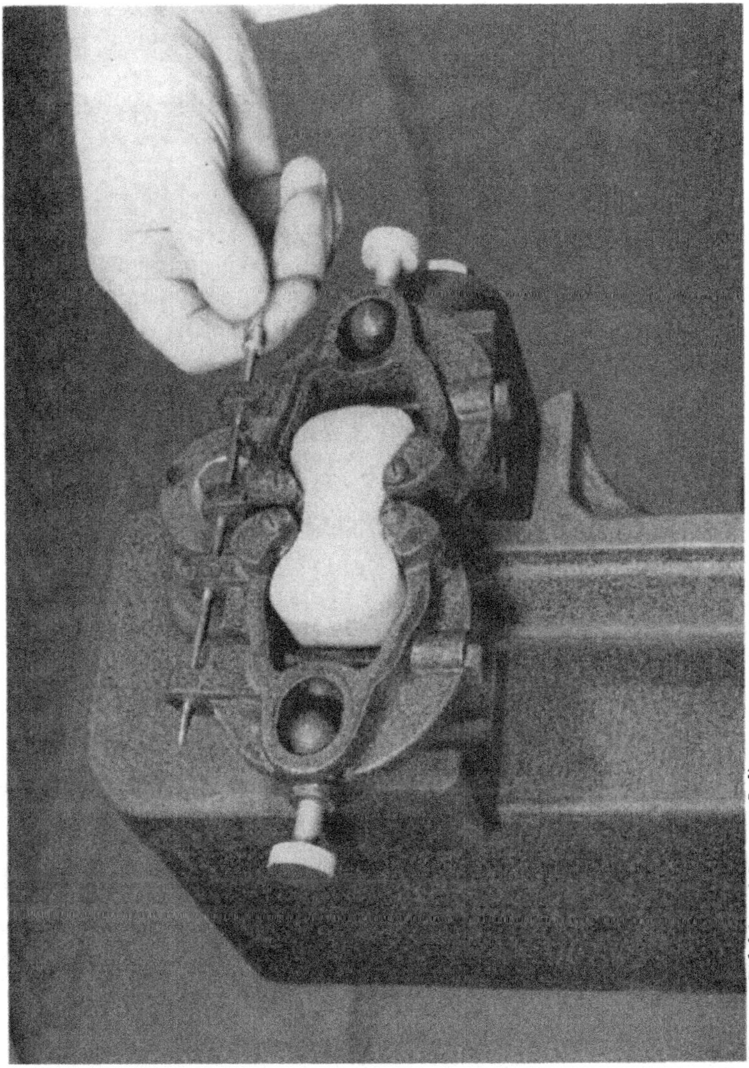

(*Courtesy Ridsdale and Company Ltd.*)

FIG. 4. Tensile core strength accessory.

where v = millilitres of air passing through the specimen in t min; h = height of the specimen in centimetres; a = cross-sectional area of the specimen in square centimetres; p = pressure of air in centimetres of water.

Routine control work is usually carried out by using a permeability meter, the test lasting a few seconds.

Shatter Index

The shatter index tester allows a standard AFS green compression specimen to fall through a height of 6 ft onto a rigid metal anvil surrounded by a $\frac{1}{2}$ in. mesh sieve and the shatter index is defined as follows:

$$\% \text{ shatter index} = \frac{\text{weight of sand retained on sieve}}{\text{total weight of AFS specimen}} \times 100 \quad (2)$$

AFS Clay Content

AFS clay may be defined as the fine material in a sand of particle size 20 μ (0·0008 in.) or less in diameter and the total clay content can be determined by using the rapid sand washer shown in Fig. 5. The method of determining the clay content is as follows:

(1) Place 50 g of sand, previously dried at 110°C, in the wash bottle.
(2) Add 475 cc of distilled water and 25 cc of 3% caustic soda solution.
(3) Agitate for 5 min with the electric stirrer.
(4) Fill up with water up to the mark on the wash-bottle glass.
(5) Stir with a glass rod for a few seconds and allow to stand for 10 min.
(6) Syphon liquid out of wash bottle.
(7) Fill up to the mark on the wash bottle with water. Stir for a few seconds with a glass rod and allow to stand for 10 min.
(8) Syphon liquid out.

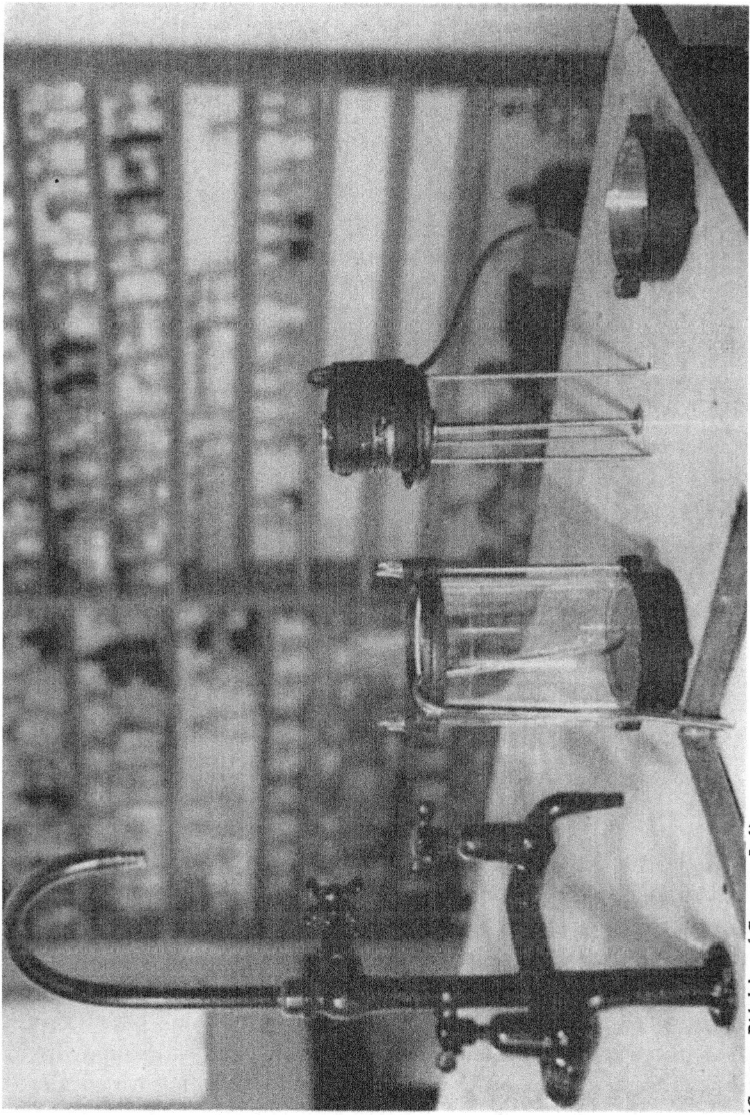

(*Courtesy Ridsdale and Company Ltd.*)

FIG. 5. Rapid sand washer.

(9) Repeat operation (7).
(10) Syphon liquid out.
(11) Fill with water up to the mark. Stir with a glass rod for a few seconds. Allow to stand for 5 min and then syphon liquid out.
(12) Continue repeating operation (11) until the liquid is clear.
(13) Syphon liquid out and detach the glass cylinder. Decant excess water from the base cup.
(14) Dry completely at a temperature of 150°C and weigh to find weight of clay-free sand W g.

The AFS clay content is then calculated as follows:

$$\% \text{ clay} = (50 - W) \times 2 \tag{3}$$

Sieve Analysis

A set of sieves of sizes as in Table 1 are placed one on top of the other with the coarsest sieve on the topmost position. The finest sieve rests on a pan, and a known weight of the clay-free dry sand is placed on the topmost sieve. The bank of sieves is shaken with a gyratory motion for 15 min. The amount of sand on each sieve is then gently tapped on a piece of paper and weighed. Graphs are plotted with "per cent retained" or "cumulative per cent retained" as ordinate against the respective sieve numbers as the abscissa.

General Notes

When testing incoming natural sands or material from the mill care should be taken that the sample is a truly representative one. AFS stipulate that for the strength and permeability measurements at least three samples should be tested and the average reported. If a sample gives reading that varies more than 10% from the average, a fourth sample must be tested. If more than three blows are used to ram specimens, this must be mentioned while reporting the results. A similar procedure is adapted for testing for shatter index, but specimens are always rammed by three blows.

TABLE 1. TESTING SIEVES

BS Specification			Boswell classification		Equivalent US series		
Mesh No.	Aperture sizes in.	Aperture sizes μ	Grade	Dia. of particles (mm)	Aperture sizes in.	Aperture sizes μ	Mesh No.
5	0·1319	3350	Gravel	Greater than 2·00	0·1320	3353	6
6	0·1102	2800			0·0937	2380	8
7	0·0945	2400					
8[1]	0·0787	2000					
10[1]	0·0661	1680	Very coarse sand	2·00 to 1·00	0·0661	1680	12
12	0·0551	1400					
14	0·0472	1200			0·0469	1191	16
16[1]	0·0394	1000					
18	0·0335	850	Coarse sand	1·00 to 0·50	0·0331	840	20
22[1]	0·0280	710					
25	0·0236	600			0·0232	590	30
30[1]	0·0197	500					
36	0·0165	420	Medium sand	0·50 to 0·25	0·0165	420	40
44[1]	0·0140	355					
52	0·0118	300			0·0117	297	50
60[1]	0·0098	250					
72	0·0083	210	Fine sand	0·25 to 0·10	0·0083	210	70
85	0·0071	180					
100[1]	0·0059	150			0·0059	149	100
120	0·0049	125					
150[1]	0·0041	105			0·0041	105	140
170	0·0035	90	Coarse silt	0·10 to 0·05			
200[1]	0·0030	75			0·0029	74	200
240	0·0026	63					
300[1]	0·0021	53			0·0021	53	270

Pan material passing No. 300 mesh } Fine silt 0·05 to 0·01. Clay grade less than 0·01.

[1] Sieves recommended for grading foundry sands.

Hardness Tester

Measurement of mould hardness, although not strictly a part of sand testing, is carried out by the sand controller, and parts of a sand-testing laboratory equipment are green and dry hardness testers. The green hardness tester is pressed squarely against the rammed mould and the hardness number read directly from the dial. Testing for the surface hardness of dry moulds or cores involves scratching the surfaces with a plough incorporated in a separate core hardness tester.

Gas Evolution Apparatus

The amount and rate of gas evolution can be measured by heating core materials in a closed silica tube. It is usual to select an arbitrary temperature, e.g. 900°C. The amount of gas evolved depends on the quantity of the volatile binder present and may be of the order of 7–20 cc/g of the core material. It is only by experimentations and shop-floor observation that the rate and total amount of gas in a core can be optimized.

An apparatus is available which incorporates a 2 in. diameter fused silica tube furnace capable of a maximum operating temperature of 1000°C. One end of the tube is closed and the other end is fitted with a water-cooled metal closure to which a spring gun is attached for ejecting the sand specimen contained in a thin-walled copper cartridge. The cartridge is ejected into the hot zone of the furnace which is sealed, and the pressure developed by the gas is measured on a manometer at regular intervals giving the rate of gas evolution. The apparatus is calibrated to convert the pressure readings into volumes.

Specific Surface

Sand grains can be round, angular or compounded. A round, that is, a completely spherical shape, is seldom found, and foundrymen have to be satisfied with near spherical grains when such sands are available. Angular sands tend to impair the flowability of mould and core materials and are brittle when compacted.

The angularity of sand grains, i.e. their deviation from a completely spherical shape, is assessed by noting the coefficient of angularity which is the ratio of S_{wa}/S_{wt} where S_{wa} is the actual specific surface (surface area per unit weight of the material) and S_{wt} is the theoretical specific surface when the particles are assumed truly spherical. Since a sphere has the least surface area, the higher the ratio of S_{wa}/S_{wt} the greater is the angularity of the sands.

An approximate value of S_{wt}, the theoretical specific surface, can be calculated from a knowledge of the sieve fractions provided the grains are equidimensional. For any spherical particle

$$S_{wt} = \frac{6}{d\rho}, \tag{4}$$

where d = the mean diameter of the particle in centimetres, and ρ = density of the material.

For silica sands, since quartz has a density of 2·65 g/cc,

$$S_{wt} = \frac{2 \cdot 26}{d}. \tag{5}$$

TABLE 2. CONVERSION FACTORS FOR CALCULATING THE VALUES OF APPROXIMATE THEORETICAL SPECIFIC SURFACE

BS Sieve No.	cm^2/g
−16+22	26·5
−22+30	38·0
−30+44	53·0
−44+60	75·0
−60+72	98·0
−72+100	124·0
−100+150	177·0
−150+200	251·0

Using eqn. (5), the theoretical specific surface for various sieve fractions are obtained by using the factors given in Table 2 as follows:

(1) Multiply the appropriate factor by the weight of sand retained on each sieve giving the theoretical specific surface for the particles on each sieve.
(2) The sum of all these surfaces divided by the total weight of the sample sieved gives the approximate theoretical surface for the sand under investigation.

Actual Specific Surface. The actual specific surface is determined by using an air permeability apparatus (Fig. 6) by the following method:

(1) Weigh out 100 g of sand and pour into the glass tube. Tap the tube until no significant reduction in the volume occupied is observed and measure the length of the column L.
(2) Switch on the pump and by means of the clamp on the bypass adjust the air pressure difference across the column to 10 cm. If the flow is too large at this pressure to read on the rotameter adjust to 5 cm pressure; if the flow is too small adjust to 20 cm pressure.
(3) When the pressure has been stabilized note the rotameter reading and convert to cc/sec (f).
(4) From the relevant curve (Fig. 7) read off the value $S_{wa}\sqrt{f}$ corresponding to the length L of the column.
(5) Calculate S_{wa} (cm²/g).

The actual specific surface can also be calculated independent of the graph in Fig. 7 by using the following expression for the BIS apparatus using quartz only:

$$S_{wa} = \frac{590}{(1-\epsilon)}\sqrt{\left(\frac{\epsilon^3 p}{fL}\right)}, \qquad (6)$$

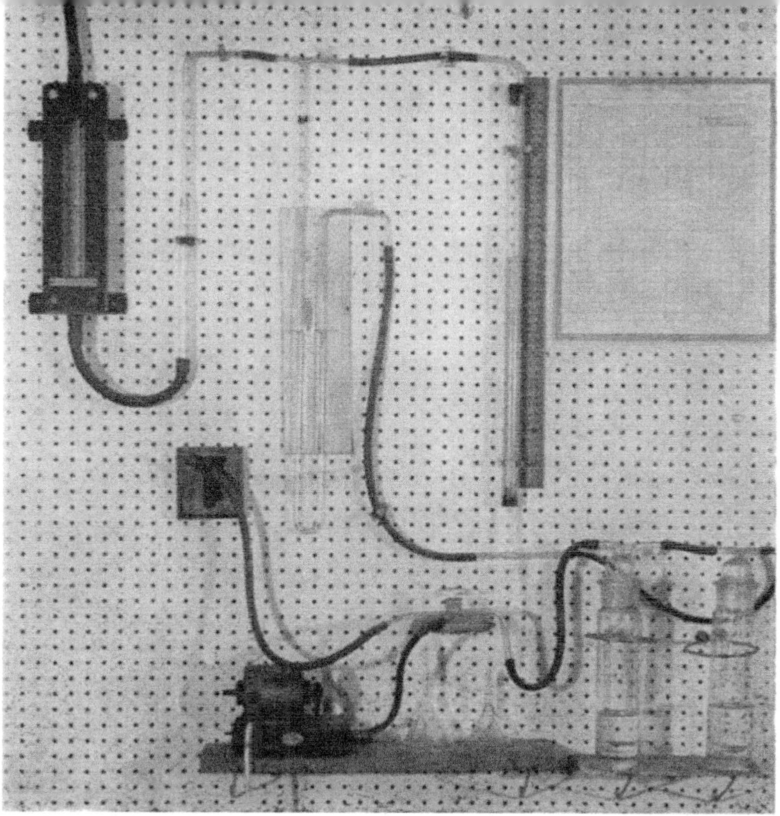

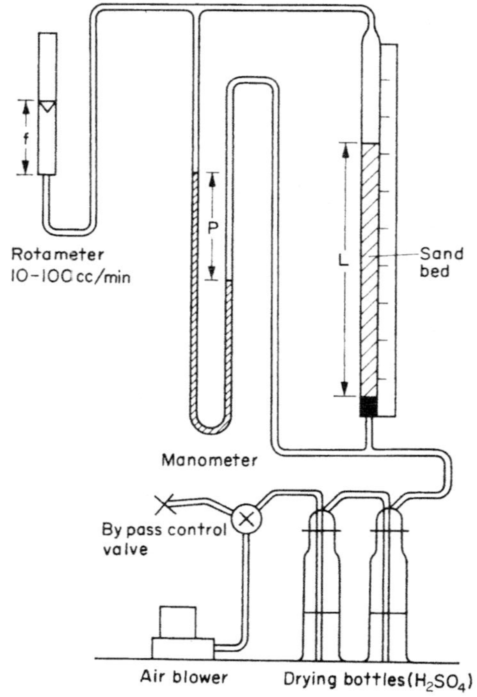

(*Courtesy British Industrial Sand Ltd.*)

FIG. 6. BIS air permeability apparatus.

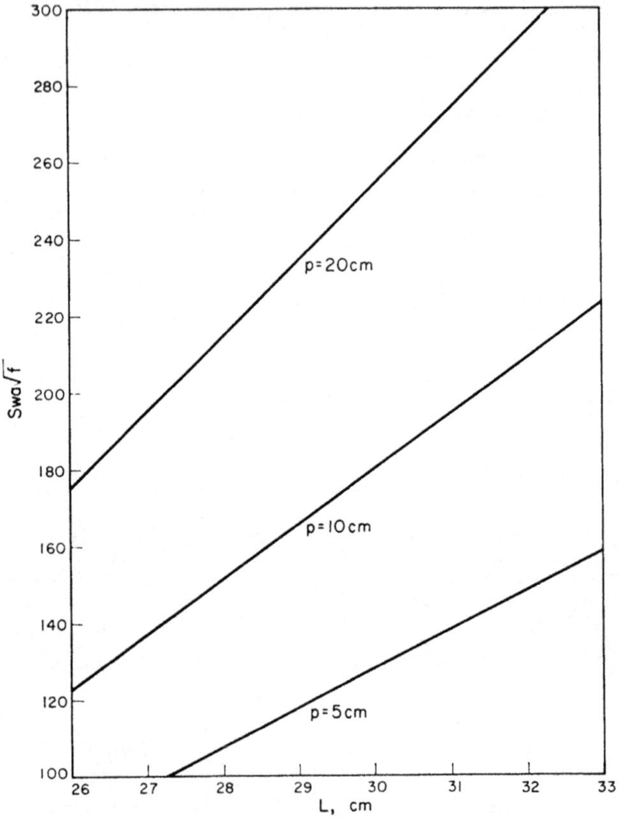

(Courtesy British Industrial Sand Ltd.)

FIG. 7. Surface area curve.

where p = pressure across the sand column in centimetres w.g., f = flow rate in cc/sec, L = length of the sand column in centimetres, and

$$\epsilon = 1 - \frac{16 \cdot 7}{L}.$$

Bibliography

Foundry Sand Handbook, American Foundrymen's Society, 1963, 7th edition.

W. Davies, *Foundry Sand Control*, the United Steel Companies Ltd., Sheffield, 1950.

CHAPTER 3

SILICA AND THE NON-SILICEOUS MATERIALS

BY FAR the most widely used basis material is silica either in synthetic mixes or in naturally bonded sands. These latter sands, for steel foundry application, need to contain at least 85% SiO_2 while the synthetic mixes employ very high silica sands containing about 98% SiO_2. There is a good supply of silica sand in most countries and it is cheaper than many other refractory materials. However, although it has a high refractoriness, the material silica has poor thermal stability. Therefore, for heavy castings at least, materials possessing superior thermal stability such as zircon and chromite have been sought. There is also some interest in exploring the possibility of using the alumino-silicates such as molochite and sillimanite for especial application in jobbing steel castings hitherto using zircon or chromite. The material olivine is finding application in Great Britain in the production of austenitic manganese steel castings because of the severe metal–mould reaction experienced with the alloy cast in green-sand mixes employing silica as the basis material.

Silica

The compound SiO_2 exists in a few allotropic modifications, the purest form being quartz. The siliceous rocks consist of quartz but the crystals are very small and are cemented together by very fine silica dust. The aggregates of these crystals are called quartzites.

Quartz undergoes polymorphous modifications upon heating.

Thus the room temperature α quartz transforms to the β form at 573°C. Quartz can be converted to tridymite upon heating at 870°C and to cristobalite at 1470°C. Quartz is liquid at 1710°C.

The various forms of quartz form a network of SiO_4 tetrahedra. Each corner of a tetrahedron shares a neighbouring tetrahedron so that each silicon atom is shared by two oxygen atoms giving the formula of silica as SiO_2. A tetrahedron is formed by each silicon cation (Si^{4+}) sharing one negative charge with each of the

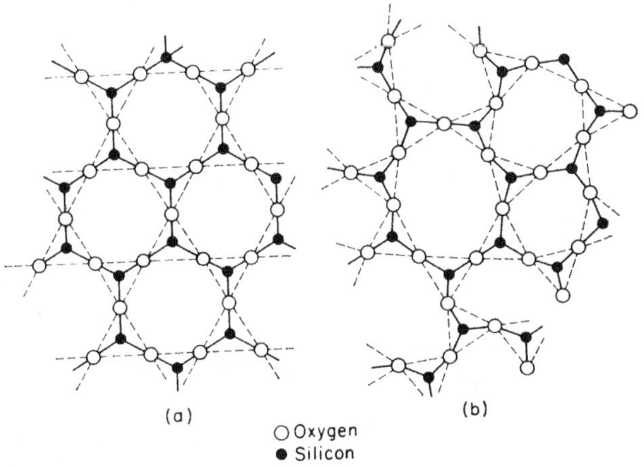

FIG. 8. Structure of pure silica (schematic). (a) The tetrahedral arrangement of crystalline silica. (b) Structure of silica in the liquid state. (*Chemistry and Industry*, 1952, p. 50.)

four oxygen ions so that the network is left with four negative charges. However, electroneutrality is possible if each oxygen ion is shared by two silicon ions from neighbouring tetrahedra giving an arrangement as shown in Fig. 8. The structures of α and β quartz are more closely packed than those of tridymite and cristobalite resulting in density differences as in Table 3.

The change of α to β quartz involves only a small alteration in bond direction and length. Consequently, the transformation

TABLE 3. DENSITY OF VARIOUS ALLOTROPIC FORMS OF SILICA

	Density (g/cc)
Quartz	2·65
Tridymite	2·26
Cristobalite	2·32

from low to high temperature quartz occurs readily, and very large and sudden expansion is noted at the transformation temperature. On the other hand, conversion to tridymite and cristobalite involves the breaking up of Si–O linkages and the rearrangement of the SiO_4 tetrahedra. This makes the reaction quartz → tridymite–cristobalite very sluggish, but can be considerably accelerated by the presence of mineralizers, e.g. Na_2O and Fe_2O_3. The cristobalite formed at high temperature does not invert on cooling, whereas the reaction $\alpha \rightleftharpoons \beta$ quartz is reversible. The complete conversion of β quartz to tridymite in the temperature range 870–1470°C is only possible if the specimen is held at that temperature for a considerable period of time. All forms of quartz undergo sudden expansion at the inversion points, and the high temperature modifications have low coefficients of expansion.

As far as the steel-foundry mould and core material is concerned, tridymite or cristobalite can only form at or near the metal–mould interface. The silica sand as used in the foundry is the low temperature quartz form and is found in many mineral deposits, in rocks such as granites, in river beds, beaches and dunes. Quartz also forms a major constituent in naturally bonded sands which contain mainly silica with a clay content varying from 10% to 20%. The clay fraction is essentially an alumino-silicate such as gibbsite ($Al_2O_3.3H_2O$). The naturally bonded sands may also contain impurities such as mica, feldspar, Fe_2O_3, etc.

Zircon

Zircon ($ZrO_2.SiO_2$) is the main constituent of igneous rocks and is also found in certain metamorphic rocks. It is associated with the heavy residues of sandstones. Concentrations of zircon with the amount of $ZrO_2.SiO_2$ varying from 20% to 40% are found abundantly in Australia, India and the USA in association with rutile, monazite and ilmenite. Zircon, as used in the steel foundry, is a by-product, for example, in the processing of monazite from the deposits, and is separated either by electrostatic method or by flotation in oil. The latter method of obtaining zircon leaves oily patches on the surfaces of the grains which result in inferior bonding properties in mould materials using zircon as the basis material.

Zircon has a high chilling power and is not wetted by molten steel. It has low expansion characteristics and requires less bond than silica sand. It has a specific gravity of the order of 4·75, and has a theoretical composition of 67·2% ZrO_2 and 32·8% SiO_2. In practice, it may contain about 1·5% Al_2O_3 with traces of impurities such as Fe_2O_3 and TiO_2. Zircon dissociates above the temperature 1550°C, but this does not appear to have any deleterious effect on the quality of steel castings.

Chromite

A South African chrome ore crushed and suitably graded is finding application in steel foundries and is believed to have very superior refractoriness. It is finding application as a facing material in cores and moulds for heavy steel castings. The chromite sand has a specific gravity of about 4·50. The thermal expansion of bentonite-bonded chromite at 1000°C is of the order of 0·007 in./in. The corresponding values for similarly bonded silica sand and zircon are 0·024 and 0·003 respectively. That is, the thermal expansion characteristic of chromite, although much more favourable than silica, is inferior to zircon. A typical chemical analysis of chromite is 45% Cr_2O_3, 25% Fe_2O_3, 10% MgO, 14% Al_2O_3, 2% SiO_2 with small amounts of CaO, MnO and TiO_2.

Olivine

Olivine is a generic term for various combinations of forsterite ($2MgO.SiO_2$) and fayalite ($2FeO.SiO_2$) in which forsterite predominates. The grade of olivine used in steel foundries in Great Britain is imported from Norway and is generally known to change colour variously upon heating up to 1500°C as noted during measurements of high temperature properties in connection with the sodium silicate–CO_2 process, and is probably the pyroxene mineral enstatite, a magnesian meta-silicate occurring as an isomorphous mixture with $FeO.SiO_2$. Enstatite inverts to clinoenstatite at 1140°C, this temperature being further lowered in the presence of $FeO.SiO_2$. Olivine has a specific gravity of about 3·3 and a lower thermal expansion than silica.

Bibliography

C. Bodsworth, *Physical Chemistry of Iron and Steel Manufacture*, Longmans, London, 1963.

J. H. Chesters, *Steelplant Refractories*, the United Steel Companies Ltd., Sheffield, 1957.

F. H. Norton, *Refractories*, McGraw-Hill, 1949, 3rd edition.

CHAPTER 4

SOURCES OF SAND

THE workability and the physical properties of mould materials are very much governed by the grain size and shape of the silica sand when this is used as the basis material, while the refractoriness of the grains will be governed by the mineralogical constitution of the sand. Sands are classified according to their stratigraphical group, i.e. according to their geological age, and this reflects the characteristics of the sands in terms of angularity, impurities, silts, etc. The main stratigraphical groups are as follows.

Recent Sands. These are found in estuaries, rivers and in dunes blown from the coast. They are unsuitable for steel foundry use because of the presence of foreign matters such as shells and fragments of coal and bricks.

Glacial. These deposits were formed by glacial action by sands from various parts being moved and redeposited elsewhere. Although angular, they are finding application in steel foundries because of their high silica content.

Lower Green Sand. These are so called because they tend to be tinted green due to the presence of hydrated potassium and iron silicate, the mineral glauconite. In certain deposits the glauconite is washed away and replaced by limonite, producing the buff-coloured sand. These sands are round to sub-angular, being very high in SiO_2 content and constitute the major source of sand for steel foundries.

Jurassic Estuarine Sands. These are naturally bonded sands with low SiO_2 content.

Triassic Sands. These contain a fair amount of limonite and are not suitable for steel foundry application.

Permian Sands. These contain an appreciable amount of lime which gives inferior refractoriness.

Carboniferous Crushed Sandstones. These sandstones are the sources of certain naturally bonded sands such as Durham rotten rock sand and are angular. They have strong bonding properties and are used in steel foundries diluted with silica sand.

British Sands

Sands from certain sources can be of satisfactory grain shape and size with the necessary refractoriness for direct application in steel foundries. Others need washing, grading, etc., before they are suitable for use as steel foundry moulding materials.

The large sand suppliers in Great Britain possess well-equipped laboratories and pilot plants, and their continuous effort in research and development has enabled supplies of sand to be available to the steel founders for specific applications. A wide range of silica sand from a British supplier is given in Tables 4–13. Their possible specific applications are indicated and photomicrographs of a few of the sands are shown in Figs. 9–13.

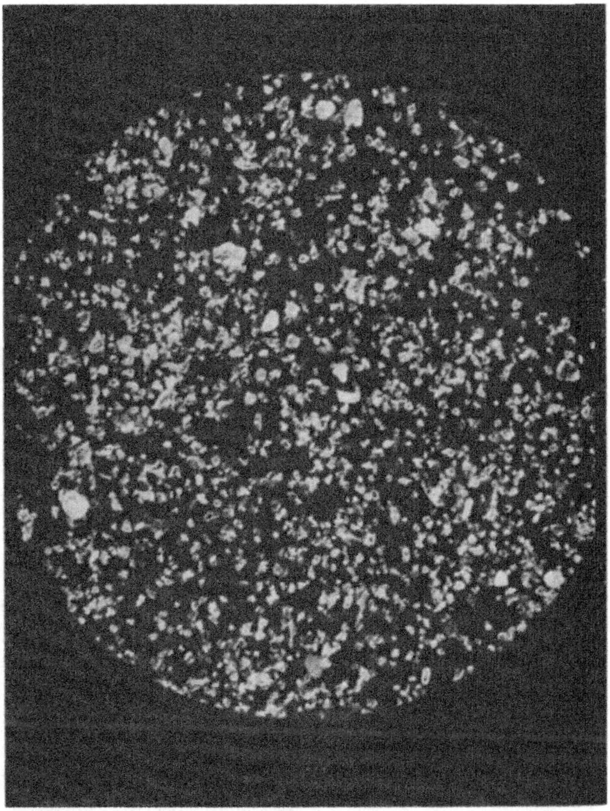

(Courtesy British Industrial Sand Ltd.)

FIG. 9. Photomicrograph of silica sand. (Redhill H—Table 13.)

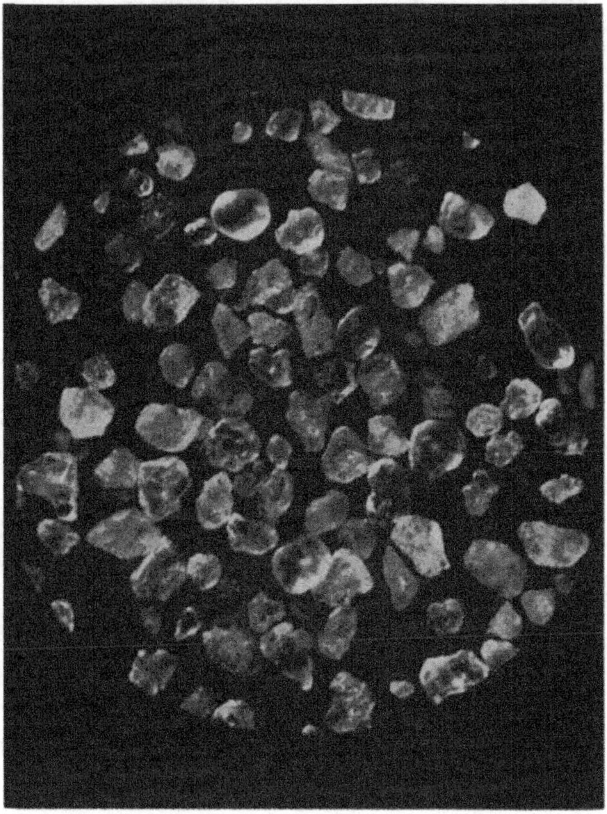

(*Courtesy British Industrial Sand Ltd.*)

FIG. 10. Photomicrograph of silica sand. (Lynn 4C—Table 6.)

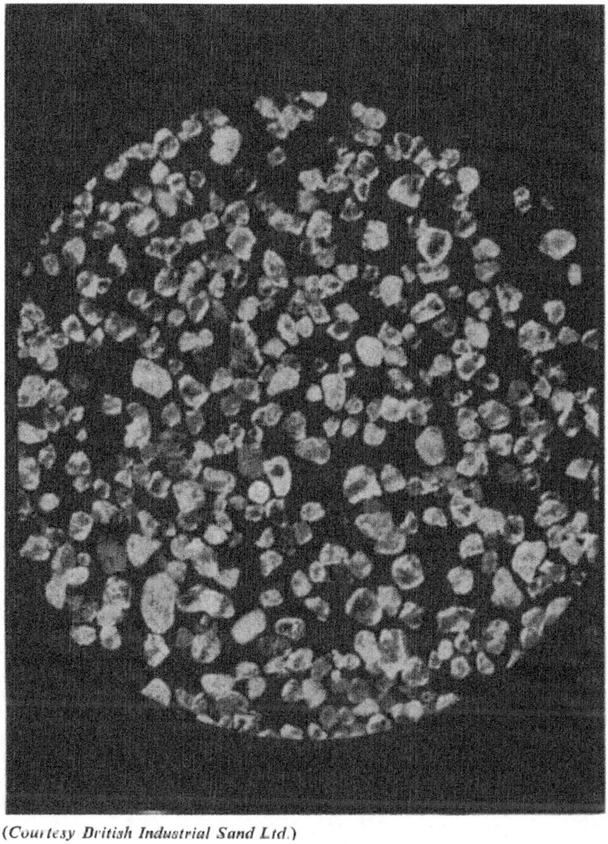

(Courtesy British Industrial Sand Ltd.)

FIG. 11. Photomicrograph of silica sand. (Redhill F—Table 12.)

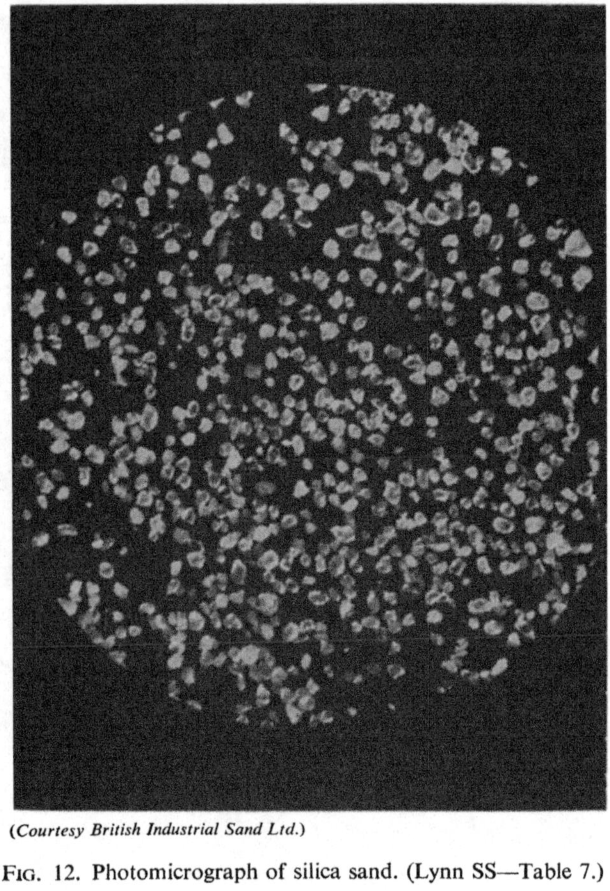

(Courtesy British Industrial Sand Ltd.)

FIG. 12. Photomicrograph of silica sand. (Lynn SS—Table 7.)

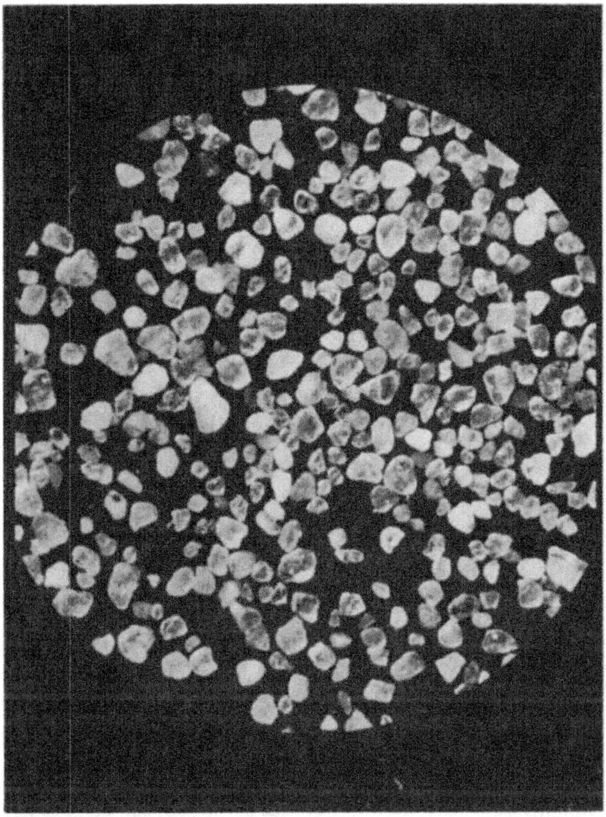

(*Courtesy British Industrial Sand Ltd.*)

FIG. 13. Photomicrograph of silica sand. (Chelford medium fine—Table 8.)

TABLE 4. PROPERTIES OF SILICA SAND
(Courtesy British Industrial Sand Ltd.)
BRERETON SAND (1964)

Silica sand: General purpose core sand.
Surface by air permeability (cm^2/g): 134.
Visual grain shape: Sub-angular to rounded.
Geological age: Glacial.
Average AFS fineness No.: 63.[1]

Average chemical analysis

	%
SiO_2	96·0
Al_2O_3	2·17
Fe_2O_3	0·16
CaO	0·08
Na_2O	0·15
K_2O	1·06
Loss on ign.	0·40
AFS clay content	0·3

Average sieve analysis

BSS No.	% retained
16	0·1
30	1·6
44	7·0
60	29·9
100	40·1
150	16·3
200	4·1
Pan	0·9

TABLE 5. PROPERTIES OF SILICA SAND
(Courtesy British Industrial Sand Ltd.)
CASTLETON FY (1964)

Washed high silica sand: General purpose silica sand for mould making.
Surface by air permeability (cm^2/g): 119.
Visual grain shape: Sub-angular to rounded.
Geological age: Moor grit.
(Base of upper estuarine series: Jurassic system.)
Average AFS fineness No.: 64.

[1] AFS grain fineness number is calculated from the sieve distribution of the sand. It is a measure of the fineness of the sand.

Average chemical analysis

	%
SiO_2	98·2
Al_2O_3	0·78
Fe_2O_3	0·35
CaO	0·03
Na_2O	0·03
K_2O	0·36
Loss on ign.	0·28

Average sieve analysis

BSS No.	% retained
16	1·3
30	13·8
44	14·3
60	26·1
100	28·6
150	10·0
200	5·9
Pan	Nil

TABLE 6. PROPERTIES OF SILICA SAND
(Courtesy British Industrial Sand Ltd.)
KING'S LYNN 4C

Washed high silica sand: Highly suitable for use as a synthetic moulding sand requiring high permeability.
Surface by air permeability (cm^2/g): 96.
Visual grain shape: Sub-angular.
Geological age: Lower Greensand.
Average AFS fineness No.: 46.

Average chemical analysis

	%
SiO_2	98·6
Al_2O_3	0·79
Fe_2O_3	0·10
CaO	0·03
Na_2O	0·03
K_2O	0·37
Loss on ign.	0·16

Average sieve analysis

BSS No.	% retained
16	nil
30	11·6
44	23·4
60	24·5
100	32·9
150	5·6
200	2·0
Pan	nil

TABLE 7. PROPERTIES OF SILICA SAND
(Courtesy British Industrial Sand Ltd.)
KING'S LYNN SS

Washed high silica sand: A fine sand giving excellent casting finish, particularly suitable for shell moulding and as a CO_2 process.
Surface by air permeability (cm^2/g): 200.
Visual grain shape: Sub-angular.
Geological age: Lower Greensand.
Average AFS fineness No.: 91.

Average chemical analysis

	%
SiO_2	98·4
Al_2O_3	0·82
Fe_2O_3	0·15
CaO	0·03
Na_2O	0·04
K_2O	0·33
Loss on ign.	0·25

Average sieve analysis

BSS No.	% retained
16	nil
30	nil
44	nil
60	3·5
100	27·4
150	59·5
200	8·3
Pan	1·3

TABLE 8. PROPERTIES OF SILICA SAND
(Courtesy British Industrial Sand Ltd.)
CHELFORD MEDIUM FINE (1964)

Washed silica sand: Of medium grading and suitable for most mould- and core-making processes.
Surface by air permeability (cm^2/g): 126.
Visual grain shape: Sub-angular.
Geological age: Glacial.
Average AFS fineness No.: 55.

Average chemical analysis

	%
SiO_2	95·9
Al_2O_3	1·92
Fe_2O_3	0·16
CaO	0·06
Na_2O	0·11
K_2O	1·09
Loss on ign.	0·38

Average sieve analysis

BSS No.	% retained
16	tr.
30	3·4
44	12·1
60	37·4
100	32·6
150	10·4
200	3·1
Pan	0·8

TABLE 9. PROPERTIES OF SILICA SAND
(Courtesy British Industrial Sand Ltd.)
CHELFORD FINE (1964)

Washed silica sand: General purpose core sand particularly where good casting finish is essential.
Surface by air permeability (cm^2/g): 230.
Visual grain shape: Sub-angular.
Geological age: Glacial.
Average AFS fineness No.: 87.

Average chemical analysis

	%
SiO_2	94·7
Al_2O_3	2·73
Fe_2O_3	0·20
CaO	0·10
Na_2O	0·23
K_2O	1·53
Loss on ign.	0·40

Average sieve analysis

BSS No.	% retained
16	nil
30	0·4
44	0·4
60	5·2
100	44·9
150	33·5
200	11·4
Pan	4·0

TABLE 10. PROPERTIES OF SILICA SAND
(Courtesy British Industrial Sand Ltd.)
MESSINGHAM FINE (1964)

Washed silica sand: General purpose core sand particularly suitable for the cold set process.
Surface by air permeability (cm^2/g): 141.
Visual grain shape: Sub-angular to rounded.
Geological age: Recent.
Average AFS fineness No.: 66.

Average chemical analysis

	%
SiO_2	93·9
Al_2O_3	2·93
Fe_2O_3	0·49
CaO	0·08
Na_2O	0·21
K_2O	1·57
Loss on ign.	0·63

Average sieve analysis

BSS No.	% retained
16	nil
30	tr.
44	0·9
60	22·8
100	55·7
150	17·5
200	2·5
Pan	0·9

TABLE 11. PROPERTIES OF SILICA SAND
(Courtesy British Industrial Sand Ltd.)
MINIMUM CORE SAND (1964)

Unwashed low clay-bearing silica sand: Of particular use in core making where a high green strength is necessary, e.g. in oil sand to prevent core sagging prior to baking.
AFS clay content: 2·0% approximately.
Visual grain shape: Sub-angular.
Geological age: Estuarine.

Average chemical analysis

	%
SiO_2	94·2
Al_2O_3	2·76
Fe_2O_3	0·95
CaO	0·06
Na_2O	0·25
K_2O	0·89
Loss on ign.	0·73

Average sieve analysis

BSS No.	% retained
16	0·3
30	0·8
44	1·2
60	6·1
100	50·7
150	36·4
200	2·0
Pan	2·5

TABLE 12. PROPERTIES OF SILICA SAND
(Courtesy British Industrial Sand Ltd.)
REDHILL F (1964)

Washed high silica sand: General purpose silica sand for mould- and core-making processes. It is particularly suitable for the shell, cold set and CO_2 process.
Surface by air permeability (cm^2/g): 123.
Visual grain shape: Sub-angular.
Geological age: Lower Greensand.
Average AFS fineness No. 123

Average chemical analysis

	%
SiO_2	99·5
Al_2O_3	0·14
Fe_2O_3	0·16
CaO	nil
Na_2O	nil
K_2O	0·08
Loss on ign.	0·14

Average sieve analysis

BSS No.	% retained
16	nil
30	5·0
44	10·6
60	24·1
100	41·6
150	10·7
200	7·2
Pan	0·8

TABLE 13. PROPERTIES OF SILICA SAND
(Courtesy British Industrial Sand Ltd.)
REDHILL H (1964)

Washed high silica sand: A fine silica sand for use where excellent casting finish is necessary. It is particularly suitable for the shell moulding process.
Surface by air permeability (cm^2/g): 298.
Visual grain shape: Sub-angular.
Geological age: Lower Greensand.
Average AFS fineness No.: 105

Average chemical analysis

	%
SiO_2	99·0
Al_2O_3	0·27
Fe_2O_3	0·12
CaO	0·01
Na_2O	0·01
K_2O	0·04
Loss on ign.	0·35

Average sieve analysis

BSS No.	% retained
16	nil
30	nil
44	nil
60	0·7
100	22·7
150	29·6
200	30·6
Pan	14·6

Two naturally bonded sands are Wear and Weatherhill with SiO_2 content of 85% and 80% respectively. They both belong to the carboniferous group.

Australian Sands

The sources of sand in the Sydney area are the Triassic sandstones and shales. The naturally bonded sands are obtained from the crushed sandstones. The silica sands used in New South Wales are the dune sands from Cronulla and Gerringong, while a few examples of the naturally bonded sands are Hornsby sandstone, Helensburg loam and Redhead loam with AFS clay content varying from 18% to 22%.

The south-eastern part of Queensland has rock deposits varying in geological age up to the recent variety. The main sources of sand are the alluvial deposits formed by weathering and redeposition of older rocks between glacial and recent times. The naturally bonded sands are, for example, Ipswich loam and Meringengden sand with AFS clay content of 13·6% and 7·4% respectively. The sources of silica sand are the dune sands of medium grade and are used as the basis material for green synthetic sands.

In south Australia, the Mount Lofty ranges provide sandstones and crushed rocks as naturally bonded sands for heavy castings. In the Adelaide plain are found deposits of glacial and recent variety which provide sands with clay content of 27·7–1·2%. The Putty sand in Launceston, Tasmania, is of uniform grain size and finds use in synthetic sands for the steel foundries. Sources of other silica sands are Frankston sand in Victoria, and Dune, Bayswater and Lake Gnangara sand in Western Australia.

Indian Sands

High silica sands are available in India containing less than 2% clay, and are sedimentary in origin which need processing. Other sources of silica sand are the dunes, estuaries, rivers and deposits of recent origin. Some of these sands need washing to remove clay silt, etc., and grading to suit steel foundry application.

The quartzitic sandstones found near Jamshedpur form a source of Indian steel foundry sands, and are obtained by crushing and grading the rocks.

The silica sands are mainly Rajmahal sand found in Santhal Parganas in the state of Bihar; Allahabad sand near Shankergarh railway station on the central railways; Jubblepore sand; Hyderabad sand; and silica sands from Vengurla, Malvan and Kudal taluks of Ratnagiri district in the state of Maharastra. Sources of silica sand are also Damodar and Barakar sands in West Bengal.

Of the naturally bonded sands are Oyaria sand in West Bengal; Batala sand in the Punjab; Bhavnagar sand in Saurashtra; Londa sand in Bombay state and Avadi sand in the state of Madras. These sands are medium to fine grained containing about 80% SiO_2 with appreciable amounts of feldspar and iron oxide.

Bibliography

H. A. Stephens, *Brit. Found.* **53**, 161 (1960).
Foundry Moulding Sands Of India, CSIR, India, 1961.
Moulding Methods and Materials, American Foundrymen's Society, 1962, 1st edition.

CHAPTER 5

THE PRINCIPAL BINDERS

THE basis grains used for steel-foundry mould and core materials are mainly quartz either in the form of silica sand or naturally bonded sands together with an amount of non-siliceous materials such as zircon, chromite and olivine. The particulate matters must incorporate an amount of binder in order to impart an optimum degree of bond strength so that moulds and cores may retain their rigidity while waiting to receive liquid metal and during casting. The four principal binders employed in the steel foundry are clay-, organic-, silicate- and resin-binder.

Clays

The clays like the silicate binder fall under the general category of inorganic bond. The principal clays used in moulding materials either as added to synthetic mixes or occurring in association with naturally bonded sands are montmorillonites and kaolinites.

The sodium bentonite such as the Wyoming variety is the widely used montmorillonitic clay in steel foundries. One advantage in using sodium montmorillonite is that optimum green and dry strengths are achieved at low amounts of clay. Of the kaolinitic clays, fireclay is an example and is a mixture of clay mineral, sand and silt. The ratio of sand to silt varies according to their geographical location and an average composition is 30% clay mineral, 42% sand and 28% silt. The bonding and refractory properties of fireclay from various sources vary widely and a very large amount of the material is needed to develop an optimum bond.

Clays are known to be very small plate-like substances of particle size in the range 0·01–100 μ containing a large proportion of Al_2O_3 and SiO_2. The particles may contain various organic matters such as lignite associated with them and also other impurities such as quartz and feldspar. Apart from these, the surfaces of the clay-flakes may have various adsorbed ions such as Na^+, Ca^{2+}, SO_3^-, etc. The impurities and the adsorbed ions affect the plasticity of a clay and hence the bond strength and workability of a mould material.

Base Exchange Capacity. These cationic and anionic particles are not part of the clay structure but are loosely held by unsatisfied bonds and, therefore, can be replaced by other ions. For example, if a clay containing an amount of calcium ions is brought into contact with Na_2CO_3, some of the Na^+ ions will attach themselves on the clay surfaces while the chemical equivalent of Ca^{2+} ions will migrate to the sodium carbonate compound. The base exchange capacity of a clay is expressed as milliequivalents per 100 g clay. The base exchange capacities of montmorillonite, kaolinite and a third clay group called illite are given in Table 14. Finer particles provide more exposed surfaces and this is probably the reason why bentonites have high base exchange capacity.

TABLE 14. BASE EXCHANGE CAPACITIES OF CLAYS

Clay group	Base exchange capacity (m-equiv. per 100 g)
Kaolinite	3–15
Montmorillonite	60–100
Illite	20–40

The base exchange capacity is utilized commercially. For example, there does not appear to be any natural source for sodium montmorillonite in Great Britain. The indigenous natural fuller's earth, a calcium montmorillonite, is treated with Na_2CO_3 or NaOH, thus effecting migration of Na^+ ions into the clay and producing sodium montmorillonite.

Gelling Index. Dry clays can develop a degree of bonding force strong enough to hold the silica or zircon grains together only when hydrated. The bond strength increases with increasing

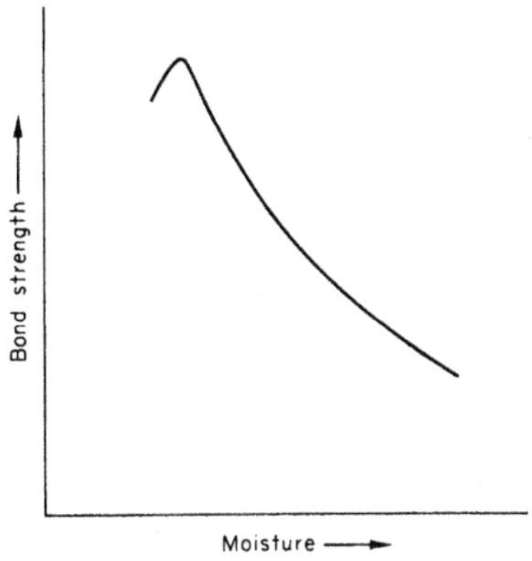

FIG. 14. Clay bond strength–moisture relationship (schematic).

amount of water reaching a maximum strength at a definite degree of hydration (Fig. 14). Thereafter, a deterioration in the bonding force occurs. For the calcium bentonites, the fall in strength occurs very sharply. The sodium bentonites can accommodate higher moisture contents and the fall in strength is more gradual. That is, the sodium bentonites are less water sensitive than the calcium variety.

Clays having the capacity to accommodate an amount of water without a sharp fall in strength are known to swell when hydrated and this is used as an index to describe the water sensitivity of clays. The intensity of swelling is established by measuring the gelling index of a clay. This is done by placing a few known weights of the clay in test tubes each containing 10 cc of distilled water. The clay–water mixes are thoroughly shaken and then allowed to stand overnight. The minimum amount of clay, w g, which will form a gel that will not flow out of the tube after being inverted next day is noted and the gelling index, GI is given by

$$GI = \frac{10}{w}. \tag{7}$$

A clay having a gelling index of less than 8·3 is regarded as non-swelling.

Liquid Limit. It has been demonstrated that the workability and strength of clays can be related to a parameter known as the liquid limit which is defined as the amount of water content of a clay at which its shear strength is so low that a groove of standard width will close upon the action of vibration. An apparatus to measure the liquid limit of a clay consists of a cup of known weight and size. This is filled with the clay–water mix to a maximum depth of $\frac{1}{2}$ in. and a groove of standard width is drawn with a special tool. The cup is then allowed to fall several times from a predetermined height by using a cam device. The percentage of water which causes the groove in the hydrated clay to close a distance of $\frac{1}{2}$ in. is defined as the liquid limit of the clay.

Liquid limits of clays can be as high as 700. Bentonites with high liquid limits are high in Na_2O content and the calcium bentonites have low liquid limits. A liquid limit higher than 500 is desirable for good workability and strength at high water contents.

pH Value. An improvement in green strength of certain naturally bonded sands which normally give an acid reaction is possible if an amount of Na_2CO_3 is added to the mill while processing the moulding material. The quantity of sodium carbonate added is normally of the order of 0·1% of the total weight of sand, and care must be taken in not adding excess sodium carbonate as this will render the sand dry and useless as a moulding material. The exact amount of sodium carbonate to be added is best established by noting the change in pH with increasing amount of sodium carbonate. Usually, a pH value of slightly over 7 but below 8 gives optimum properties.

Green and Dry Properties. pH value *per se* has not proved to be a reliable test in predicting the subsequent behaviour of a moulding material because the acidity of mixes may be affected by trace elements which may not have much bearing on the properties of compacted moulds and cores. Steel foundries often assess the suitability of a new clay or a naturally bonded sand by simply testing for green and dry properties together with refractoriness.

For example, for green sand mixes, a 5% addition of the clay is made to a stock silica sand with 3% water for a montmorillonitic clay. The clay is regarded as satisfactory if the green compressive strength is of the order of 5–7 lb/in^2 with a shatter index of about 70%. The dry compressive strengths on samples dried at 110°C for 2 hr should exceed 60 lb/in^2.

Refractoriness. The pyrometric cone equivalents of clays as used in the refractory industry are not satisfactory for describing the refractoriness of steel foundry mould and core materials. The reason is that clays such as Wyoming bentonite or a sodium silicate bond have melting points well below the metal/mould interface temperatures of steel castings. Further, the organic binders are volatile.

A phenomenon of a quartz–clay mix is that, at high temperatures, the quartz grains will expand by transforming to cristobalite/tridymite and the clay together with any low melting point constituents will melt and shrink upon cooling. The net effect is that a compacted material will show a volume change upon cooling decided by the algebraic sum of the expansion and contraction. If the volume change is positive, the indication is that shrinkage due to melting of the clay and other constituents is not so serious and the material is regarded as refractory.

The testing procedure is to ram a standard AFS compressive specimen and dry it at 110°C for 2 hr. The diameter and the height of the specimen is then measured to the nearest $\frac{1}{32}$ in. by means of a pair of calipers. The dry sample is then inserted in a cold furnace which is then heated at the rate of 20°C/min. A molybdenum wound electric furnace may be used, but a propane gas-fired furnace is quite satisfactory. The specimen is soaked at 1550°C for 2 hr and allowed to cool in the furnace. After cooling, the height and diameter of the specimen are again noted and from the initial readings on the dry sample, the per cent linear change is calculated. The appearance of the specimen after heating, that is whether the sample has glazed or fritted, is also noted.

Organic Binders

The organic binders can be classified into three main groups, viz. oil-, cereal- and resin-bond. Use of oil and cereal bonds is, perhaps, as old as the foundry industry itself, and in recent years the resins have been used very extensively in such processes as shell moulding and mould and core making by the air-set and hot-box processes.

Oil Bond. All the three types of oil, viz. vegetable oil such as linseed oil, animal oil and mineral oil as, for example, the polymers of petroleum have found application in the steel foundry.

Oils can be categorized into the air-drying and oven-drying types. Linseed oil is the latter variety and, prior to the introduction of air-drying oils, found universal application for core production. The advantage in the use of oil is that it forms a very strong bond by oxidation and polymerization, thereby producing very strong cores with hard non-friable surfaces. The other main advantage of oil sands or any core and mould material containing organic bonds is that after casting the bonding medium volatilizes giving very good breakdown of cores and moulds during knockout of castings.

The high strengths in oil sands are developed due to the presence of linoleic and linolenic acids in oils. Linseed oil is rich in linolenic acid and is the rapid drying type as opposed to linoleic acid which is slow drying.

A good test to evaluate the suitability of an oil in terms of baked strength and surface hardness of cores is to find the saponification number which is the number of milligrams of KOH necessary for the complete hydrolysis of 1 g of oil. Raw linseed oil has a saponification number of 190 and a value exceeding 130 is known to produce very satisfactory cores for steel-foundry application.

Animal oils have also been used but about 30% more oil is necessary to develop strengths comparable to vegetable oils. The basis for quick-drying oils is the mineral oils such as the petroleum polymers. Mixed oils or linseed oils containing accelerators in the form of metallic salts are manufactured as proprietary binders and find application in steel foundries.

Oils, whether vegetable, mineral or animal, have little green strength and therefore stripping of cores from core boxes becomes very difficult. Inadequate green strength also results in sagging of cores on drying plates during transportation to ovens for baking. This problem is overcome by imparting an amount of green strength to oil-bonded sands, which is done by incorporating another organic binder in the form of a cereal. Cereal, when mixed with an amount of water, imparts a degree of green bond sufficient for the cores to be stripped from the core boxes and to retain their shape until strength and rigidity are developed upon baking in the drying stoves.

Green and dry strengths in oil-bonded core materials are not compatible and, therefore, there is always some loss of baked strength when an amount of cereal addition is made. Some founders prefer the proprietary binders which are in the form of semi-solids and appear to be essentially oil-impregnated cereals.

The advantage is that addition to the mill during sand preparation is simplified.

Cereal Binders. The two main groups of cereals well proven in the steel foundry are starch and dextrin. Starch as derived from maize is not used in its raw form but is converted to its sticky consistency upon hydration by pre-gelatinizing.

Gelatinizing is carried out by passing the starch through hot rolls heated by steam.

Mould and core materials cannot be produced by using cereal binders only because the amount of green strength obtainable is not adequate. Moreover, during the flow of molten steel over mould and core surfaces, the cereals would burn freely resulting in depletion of bond. As metal flows severe mould erosion will occur and, in extreme cases, a complete collapsing of cores or moulds will take place before the casting has at least formed a hard solid skin.

The role of cereals on imparting an amount of green strength in oil sands has been mentioned. Cereals are used because they are less harmful than clays from the point of view of baked strength in oil sand cores. However, the green strength developed is very low, being of the order of 1 lb/in^2, and some sagging may occur in large cores. Starch is capable of imparting a higher green strength than dextrin for the same amount of moisture. Dextrin, on the other hand, migrates to the core surfaces which dry in air giving a hard skin, and this is known to minimize sagging. However, while using dextrin, the moisture in the sand mix should be kept low and not preferably be allowed to exceed 1%.

Apart from its use in oil-bonded sands dextrin is added to clay-bonded green sand mixes to form a hard non-friable skin on the moulds. Presence of starch decreases the scabbing propensity of green sand moulds but increases surface friability. For these reasons, both starch and dextrin are used as mixed additives in steel foundry moulding materials. Starch is also known to increase dry strengths of moulding materials.

Sulphite Lye and Molasses. Sulphite lye and molasses are water soluble and are used as additives in moulding sands. They increase dry strength and tend to migrate to the surfaces giving high surface hardness to dry moulds. Hard surfaces mean a lack of friability in the exposed mould areas and consequently mould erosion during metal flow is minimized. The proprietary core creams are mixtures blended from oils, starch, sulphite lye, etc.

Air-setting Oils. Apart from decreasing the baked strength in oil-sand cores, a clay addition impairs the blowability of core-shooting sands and a deterioration in breakdown efficiency during knockout of castings occurs. The next step in the development of oil sands has been to use air-drying oils which would impart an amount of strength to the cores if the rammed material is allowed to stand in the core boxes for a while before stripping. The oils which harden in air are referred to as air-setting oils and are produced by treating drying oils with accelerators. The amount of accelerator added controls the air-setting property and the subsequent drying time of the oil. The most widely used accelerator is sodium perborate although organic peroxides or calcium hypochlorite are also used. The basis material must be clay-free silica sand with a minimum amount of fines. The air-set core materials have excellent flowability and require about half the drying time of mixes made with raw vegetable oil. The cores strip very easily and good dimensionally accurate castings are produced.

Air-setting oils which can tolerate an amount of clay to impart a degree of green strength have also been developed and are used for both mould- and core-making purposes. An amount of green strength cuts down the stripping time after completion of ramming and frees the core box for the production of a second core. The oils which can accommodate an amount of clay are a mixture of linseed oil, synthetic resin and a dryer such as lead or cobalt naphthenate. The clay needed may be added to the silica sand–oil mix in the mill in the form of ball clay or bentonite. Alternatively, an amount of naturally bonded sand may be added to the silica sand–oil mix in the mill.

Air-setting oils have been used to prepare moulds producing castings weighing up to 10 tons. However, certain disadvantages of the process are that even a trace amount of moisture in the mix may cause mould erosion defects and the mould–core materials produced are expensive because of the high cost of both the oil and silica sand. Further, the binder, because of the presence of resin, may be thermoplastic causing cores and moulds to sag during drying.

The next logical step in the development of binders was to use the furane binders which attain optimum strengths by simply standing in air while the rammed core material is still in the core box. These binders, together with the materials used for shell mould and core making and the silicate-bonded sands, are discussed in separate chapters.

Bibliography

J. M. Middleton, *Brit. Found.* **58** (1), 13 (1965).

H. W. Dietert, *Foundry Core Practice*, American Foundrymen's Society, 1950.

Moulding Methods and Materials, American Foundrymen's Society, 1962, 1st edition.

J. M. Navarro and H. F. Taylor, *Brit. Found.* **52** (8), 342 (1959).

CHAPTER 6

CLAY BOND

CLAY is defined as an earthy material showing marked plasticity and is composed essentially of Al_2O_3 and SiO_2. It is an agglomerate of minute plate-like particles called clay minerals within a size range of $0 \cdot 01$–$100\,\mu$. The plasticity of clays is markedly affected by cations Ca^{2+}, Na^+, etc., or anions such as SO_3^- which attach themselves to the surfaces of the plates. The plasticity is also affected by the presence of impurities, viz. quartz, mica or feldspar. These impurities also have pronounced fluxing action at high temperatures resulting in a lower refractoriness of the clay.

Clays are derived from the weathering of rocks rich in feldspar. The process is a gradual removal of alkalis and silica. The clays found in the same part as the source from which they were formed are called residual clays and need washing because of the presence of pieces of the original rock. The main sources of foundry clays are deposits from lakes, oceans or swamps.

Clay Types

The structural configuration of clays belong to the layer lattice group of minerals in which sheets of oxygen ions O^{2-} are laid each on top of another. The spaces between these sheets are filled with cations such as Al^{3+}, Si^{4+}, etc.

There are three main groups of clay all of which find application in providing the bonding medium for steel foundry moulding materials, viz. kaolinite, illite and montmorillonite.

Kaolinite. Kaolinite in the form of fireclay, flint clay and china clay conforms to the general formula $Al_2O_3.2SiO_2.2H_2O$ and has a theoretical Al_2O_3 content of about 39·3% and 14% of combined H_2O. The alumina content in some clays increases because of the presence of aluminium hydroxide in the form of gibbsites, bauxites and diaspores or dehydrated alumino silicates such as sillimanites, kyanites or andalusites.

Illite. These are represented by the chemical composition

$$(OH)_4K_y(Al_4.Fe_4.Mg_4.Mg_6)(Si_{(8-y)}.Al_y)O_{20},$$

and they derive from weathering of minute mica, sericite and biotite flakes. These clays are found abundantly in most natural sands such as Durham rotten rock moulding sand. Some naturally bonded sands also contain kaolinite or limonite ($Fe_2O_3.H_2O$). For example, the red moulding sands contain a fair amount of limonite.

Montmorillonite. Montmorillonites conform to the general formula $Al_2O_3.4SiO_2.xH_2O$ and are derived from the mechanical and chemical weathering of volcanic lava. The alumina is partially replaced by MgO, CaO, Fe and sometimes alkalis.

The most common montmorillonitic clays used in the steel foundry are sodium and calcium bentonites. The essential difference between the two types is the presence of calcium or sodium ions between the silicon–oxygen layers of the montmorillonite crystal.

The Layer Sequence. The layer sequence of the three clays is shown schematically in Fig. 15. In the kaolinite group, the stacking is alternate layers of silica (SiO_2) and gibbsite ($Al(OH)_3$). In the illite and montmorillonite groups a gibbsite layer is sandwiched between two silica layers.

In the montmorillonite group some of the Al^{3+} ions are replaced by an equal number of Mg^{2+} ions. The excess negative charges are

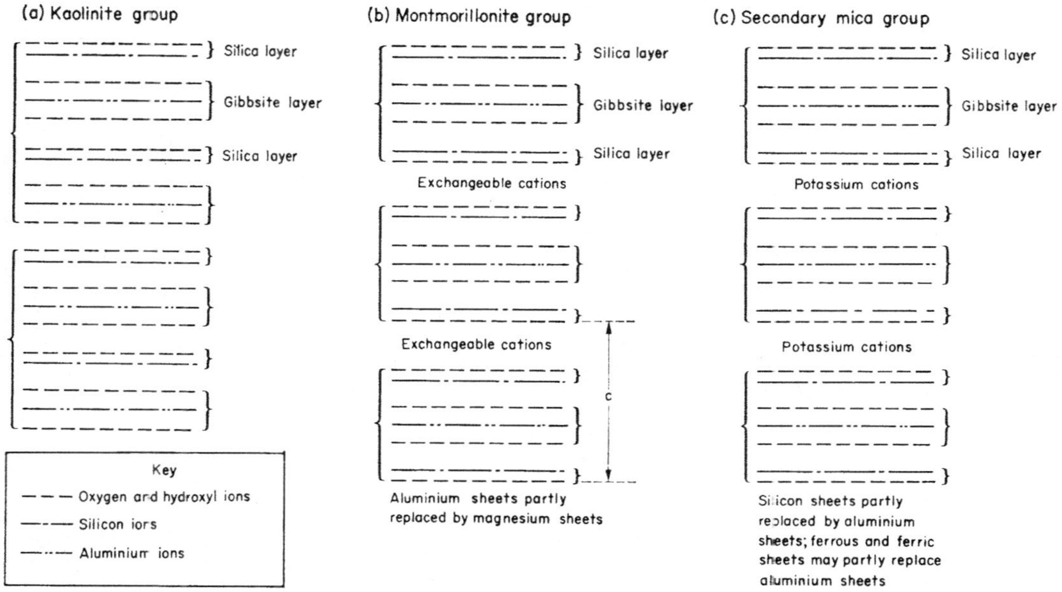

FIG. 15. Layer sequence in clay minerals (Davies).

balanced by Na^+, Ca^{2+} or other cations loosely held between the units, thus producing sodium or calcium montmorillonites such as Wyoming bentonite and natural fuller's earth respectively. The cations are associated with water molecules and the distance *c* (Fig. 15) may vary from 9·6 Å to 21·4 Å units. The montmorillonites have an expanding lattice and are termed the swelling type clays. In the illite group one in every four Si^{4+} ions in the silica layer is replaced by one Al^{3+} ion. The excess negative charge is balanced by potassium ions and the linkage is stronger than in the montmorillonites. Consequently, water cannot enter the lattice and illites like kaolinites are non-swelling.

Green Bond

The mechanism of clay bond is described in terms of a net intermiceller force at the quartz–clay and clay–clay interfaces which is initiated by the preferential adsorption of cations and anions on hydrated clay surfaces.

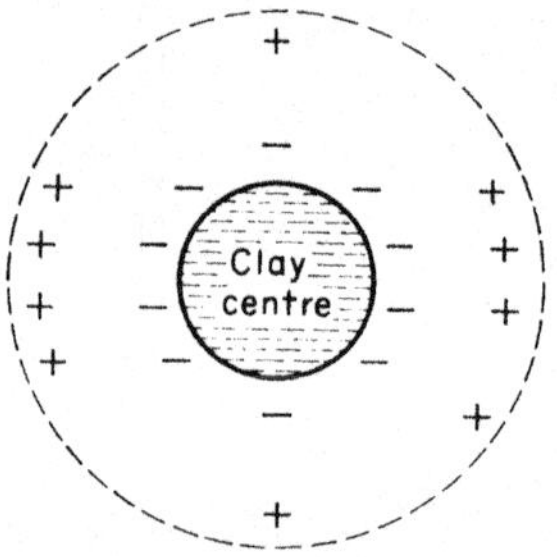

FIG. 16. A clay micelle.

Kaolinites contain broken valence bonds on their surfaces giving rise to active spots on the crystals. Upon adding water to a dry kaolinite, the OH^- ions are preferentially adsorbed on the clay centre and form an integral part of the crystal. The positive ions are attracted by the OH^- ions but repelled by the clay centre resulting in the ions taking up equilibrium positions. The cations take up a position as shown by the dotted line in Fig. 16 which is the position where the electrostatic force tends to be zero and a clay micelle is produced. The magnitude of the repulsive forces is defined by the type of cations such as Na^{+}, H^{+} and Ca^{2+} and the combined effect of the attractive and repulsive forces is to have the tendency of neutralizing the charge on the clay particle. In a mould material, the clay and the quartz particles are dispersed in a water medium. The water forms neutralized clay micelles and their kinetic energy causes them to move towards one another. There occurs a force of attraction between the unlike ions and a force of repulsion between the cations themselves and the clay centres. The force of attraction increases and that of repulsion

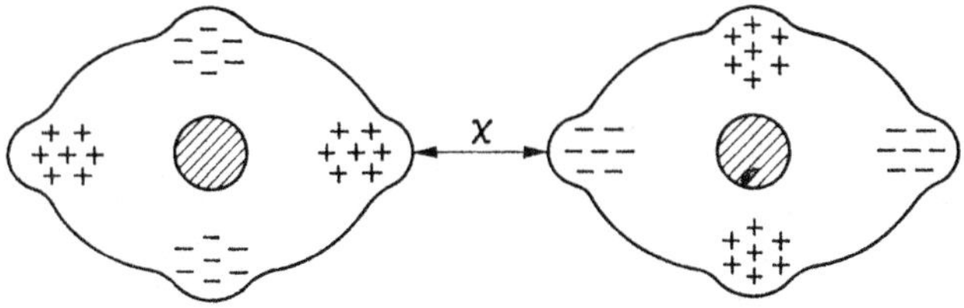

(*Courtesy American Foundrymen's Society.*)

FIG. 17. Clay dipole.

decreases as the distance between the clay micelles increases resulting in a net intermiceller force.

The drawing together of two micelles will cause orientation of unlike ions forming a clay dipole (Fig. 17) and a maximum attractive force is at an optimum distance of separation x. There will be many such dipoles in a clay–water medium which will be oriented selectively forming a network binding the hydrated clay particles together.

Depending on the type of clay, that is on the magnitude of charge on its broken surface, a maximum degree of hydration will be necessary to develop a dipole completely. This is why strengths of clay-bonded sands increase with increasing amount of water addition to a maximum value. As the water addition is increased further it enters the spaces between the dipoles causing a

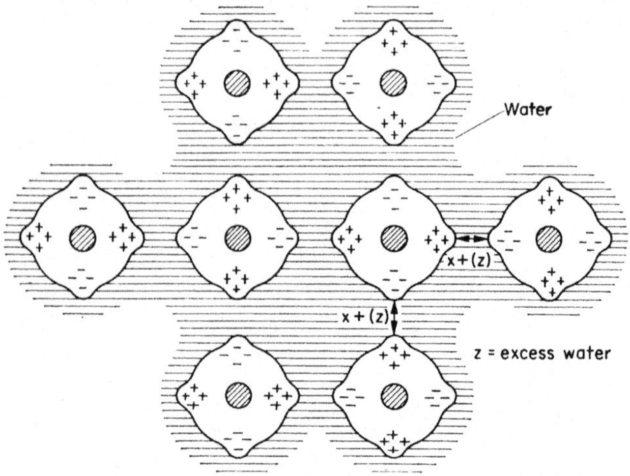

(*Courtesy American Foundrymen's Society.*)

FIG. 18. Clay dipoles in excess water.

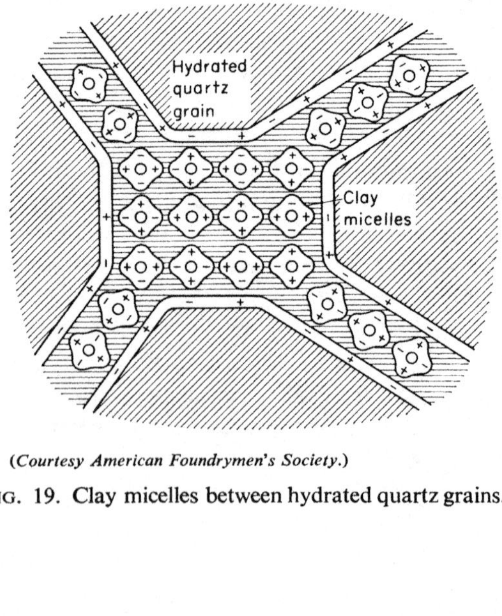

(*Courtesy American Foundrymen's Society.*)

FIG. 19. Clay micelles between hydrated quartz grains.

separation of the dipoles to a distance greater than x (Fig. 18) resulting in a decrease in the net intermiceller force.

Quartz–Clay Interface. Quartz surfaces also have broken valence bonds and consequently form hydrated quartz micelles. The clay dipoles surrounding adjacent quartz grains are held on the sand surfaces (Fig. 19) although the force at the clay–quartz interface is weaker than the intermiceller bonding between the clay dipoles.

Dry Bond

In a green sand, although the strength increases with increasing amount of water to a maximum, in the normal range of moisture in foundry sands compatible with the workability of the moulding material, the distance between clay dipoles is always greater than the critical value x because of the presence of excess water between the intermiceller spacings. However, if the clay–water–quartz compacts are heated sufficiently to allow for moisture evaporation, the dipoles are drawn together and there is a volume decrease. Shrinkage increases with time until the critical distance x is achieved between the dipoles. Further drying has little effect on the magnitude of dry strength once the critical amount of moisture has been reached.

Moulds and cores are dried in stoves whose temperature must be critically controlled. A temperature above 300°C of the stoves is detrimental because of the changes occurring in clay structures at higher temperatures. A minimum drying time must be employed to achieve the maximum dry bond.

It has been shown that the dehydration reaction in a quartz–clay–water system is a first-order reaction given by

$$\log_e \frac{L\infty - L}{L\infty} = -kt, \tag{8}$$

where L = loss of water in time t; $L\infty$ = loss of water when the dehydration reaction is complete at the temperature; $k =$ constant.

Green and Dry Bond in Various Clays. The mechanism of green and dry bonds in montmorillonite, kaolinite or illite is governed by the same phenomenon although because of its swelling nature Na-montmorillonites form complex micelles. It is known that up to four water molecules can enter the space between the silicon–oxygen layer in sodium bentonite resulting in a loss of cohesion in the clay lattice. The effect is that these clays fail under low shearing forces giving a lower green bond than the non-swelling calcium montmorillonite. Green bond is also improved by small ionic size which when associated with a low degree of hydration favours a thin double diffuse layer with low repulsive forces between clay micelles.

The advantage of the high degree of hydration is that the sodium montmorillonites such as Wyoming bentonite are spread around the quartz grains as a coherent thread-like network. Upon drying, as the dipoles come closer together, a continuous bond envelope resists shearing forces more than a calcium montmorillonite such as fuller's earth which gives lower dry strengths than sodium bentonite.

The kaolinites and illites give inferior bond strengths to the bentonites. This is attributed to the larger particle size, that is smaller surface area, of the former two clays.

Effect of Heat on Clays

The changes occurring in hydrated clays upon heating to their liquefaction temperatures can be followed variously by measuring loss in weights or by differential thermal analyses and X-ray methods. The main changes which occur are:

(1) Loss of adsorbed moisture from the surfaces of the clay plates and also from between the Si–O layers.
(2) Loss of lattice or constitutional water.

(3) Crystallization of alumina followed by conversion of silica to cristobalite and the ultimate combination of these to form mullite ($3Al_2O_3.2SiO_2$). For example, in the temperature range 850–1050°C kaolinite dissociates into mullite and tridymite; mullite crystallizes above 1350°C and tridymite converts to cristobalite at 1470°C. The eutectic containing 5·5% Al_2O_3 and 94·5% SiO_2 forms at 1545°C.

(4) Fusion and liquefaction of the clay. For example, kaolinite melts at a temperature of about 1750°C.

Of these loss of both adsorbed and constitutional water is of direct practical importance to the foundry. Adsorbed water controls both the green and dry compressive strengths and workability of mould or core materials. Loss of constitutional water means that the clay has lost its layer lattice structure and is non-bonding. This gives rise to dead clay in cast moulds and necessitates addition of fresh clay to compensate for loss of bond strength when the sand from the knockout system is re-used for moulding and core-making purposes.

Adsorbed Water. For the non-swelling group such as kaolinite loss of adsorbed water involves only those attached to the external surfaces of the clay plates. In montmorillonites the excess water which has entered the lattice is also involved. Adsorbed water from foundry clays can usually be removed in the temperature range 110–150°C.

Constitutional Water. Generally, the constitutional water is lost over a range of temperature which for the three foundry clays is as follows:

Kaolinite	450–550°C
Illite	350–550°C
Montmorillonite	550–650°C

The factors affecting the loss of lattice water appear to be the particle size, chemical composition and the structure of the clay.

Bibliography

L. M. DIRAN and H. F. TAYLOR, *Trans. AFS*, **60**, 356 (1952).
R. K. SCHOFIELD, *Trans. Brit. Ceram. Soc.* **39**, 147 (1939–40).
P. MURRAY and J. WHITE, *Ibid.* **48**, 187 (1948–49).
A. L. JOHNSON and F. H. NORTON, *Jl. Amer. Ceram. Soc.* **24**, 189 (1941).
F. H. NORTON, *Refractories*, McGraw-Hill, 1949.
J. J. MARAIS, *Brit. Found.* **52**, 4 (1959).
R. E. GRIM and W. F. BRADLEY, *Jl. Amer. Ceram. Soc.* **23**, 242 (1940).
S. DAVIDSON and J. WHITE, *Foundry Trade Jl.* 165, 235 (August 1953).

CHAPTER 7

SAND PREPARATION

FOUNDRY mould materials can be broadly classified into synthetic and naturally bonded sands. A synthetic sand may be defined as one where the bonding material is added in a predetermined quantity to clean filler grains such as silica sand or zircon, chromite, etc., where non-siliceous materials are used. The bulk of the synthetic sands, silicate- or resin-bonded and air-set mould materials employ silica sand as the basis material and the clay-bonded synthetic mixes usually incorporate Wyoming bentonite together with cereal binders and moisture. The naturally bonded sands as received from the sand quarry contain an adequate amount of clay in the form of, for example, kaolinite and can be used directly as a moulding material. However, since there is no control in the amount of clay present, certain natural sands need to be diluted with silica sand. The latter addition also improves the refractoriness of the material.

Sand Mixes

The first step in sand preparation is to formulate the mixes to give properties adequate to produce a casting free from defects. The mixes must be economically feasible and only where necessary, for example, in heavy sections should expensive materials be used. Some typical production sand mixes are shown in Table 15 with their average green and dry properties. The weight required to prepare a standard AFS compressive specimen is also given and indicates how the bulk densities of the mould or core materials vary for similar ramming energies.

TABLE 15. PRODUCTION SAND MIXES

Mix 1. Dry sand (1500 lb mix)

	%	Properties[1]	
Silica sand	69	M	5·6
Natural sand	25	GP	113
Sodium bentonite	2·5	GC	5·6
Fireclay	2·5	SI	90
Molasses	1·0	DP	175
		DC	195
		Wt. AFS	190

[1] Abbreviations used for properties: M = % moisture. GP = green perm. No. GC = green compressive strength (lb/in²). SI = shatter index (%). DP = dry perm. No. DC = dry compressive strength (lb/in²). Wt. AFS = weight of standard AFS compressive sample (g). DT = dry tensile strength (lb/in²).

Mix 2. Core sand (large jobs) (1000 lb mix)

	%	Properties	
		M	3·4
Silica sand	67	GP	109
Natural sand	30	GC	2·7
Starch	1	DP	180
Core cream	2	DC	1000
		DT	205
		Wt. AFS	175

Mix 3. Core sand (small jobs) (1000 lb mix)

	%	Properties	
		M	4·2
Silica sand	82	GP	111
Natural sand	14	GC	2·8
Vegetable oil	1	DP	159
Core cream	2·75	DC	1040
Air-drying oil	0·25	DT	160
		Wt. AFS	175

Mix 4. Weak core sand (1500 lb mix)

	%	Properties	
		M	6·8
Silica sand	31	GP	72
Natural sand A	31	GC	6·9
Natural sand B	31	DP	140
Crushed shell moulds	4	DC	186
Ball clay	1	DT	25
Core cream	2	Wt. AFS	190

TABLE 15. (*contd.*)

Mix 5. Core shooter sand (700 lb mix)	%	Properties	
Silica sand	95		
Natural sand	0·5		
Ball clay	1·5	(not determined)	
Vegetable oil	3·0		

Mix 6. Green facing sand (2500 lb mix)	%	Properties	
		M	3·1
Silica sand	94·25	GP	218
Wyoming bentonite	5·00	GC	5·1
Starch	0·75	SI	82
		Wt. AFS	160

Mix 7. Backing sand	%	Properties	
Reconditioned sand	99·5	M	3·1
Ball clay	0·5	GP	209
		GC	5·6
		DC	72
		SI	74

Mix 8. Backing sand	%	Properties	
		M	4·3
Reconditioned sand	99·5	GP	137
Wyoming bentonite	0·5	GC	8·3
		DC	90
		SI	75
		Wt. AFS	170

Mix 9. Unit sand (500 lb mix)	%	Properties	
		M	2·5
Return sand	80	GP	188
Silica sand	19·25	GC	7·8
Wyoming bentonite	0·5	Wt. AFS	160
Starch	0·25		

The dry-sand mould material (mix 1) is used for heavy steel castings, using about 70% silica sand and 30% of a naturally bonded sand. An all silica sand mix is not used for reasons of economy and the purpose of adding molasses is to give hard mould surfaces upon drying, and the role of fireclay is to improve the dry and hot properties. Use of bentonite allows high dry strengths to be developed at low moisture content. Even in dry-sand mixes, moisture should be kept as low as possible if the flowability of the material is not to be unduly sacrificed. A mixed clay bond such as mix 1 in Table 15 is quite commonly used for dry-sand moulding. The green facing sand (mix 6) incorporates an amount of starch which minimizes scab defect. However, if the surfaces of the moulds are excessively friable an amount of dextrin should also be incorporated. Alternatively, the green mould surfaces should be sprayed with a spirit based air-setting resin solution. The backing sand (mixes 7 and 8) is made up with return sand from the foundry knockout areas after reconditioning. An amount of ball clay or bentonite is added to compensate for the dead clay which results at and near the metal–mould interface because of the loss of constitutional water at high temperatures.

The green-sand facing mix (mix 6) can be used for machine moulding in conjunction with a backing sand. However, in mechanized foundries there is a preference for using unit sand (mix 9) which uses about 80% return sand from the knockout and 20% silica sand. The moisture is kept at a low value of about 2·5% to ensure that the mould material possesses good flowability and does not clog the sand hoppers, etc.

The core sands are bonded with oils and proprietary core creams. Often, for small jobs with complex re-entrant angles, baked strength of the core material is less important than its collapsibility during and after solidification of the casting both to prevent hot tear defect and to allow subsequent ease of decoring during knockout. This can be achieved, for example, by using about 4% crushed scrap shell moulds (mix 4) where the process is used in the foundry. Generally, however, core sands must be formulated to develop high dry compressive and tensile strengths with hard non-friable surfaces. Core materials based on the chemically hardened sands such as the air-set material are discussed in a separate chapter.

Sand Mills

The actual mixing operation of the basis grains, binder, moisture and additives are carried out in sand mills, and the foundry industry uses either or both a roller and impeller type mill.

Roller Mill. A roller type mill constitutes essentially a mixing pan together with one or more ploughs. The pan is filled with sand, bond, water and additives and the rollers are switched

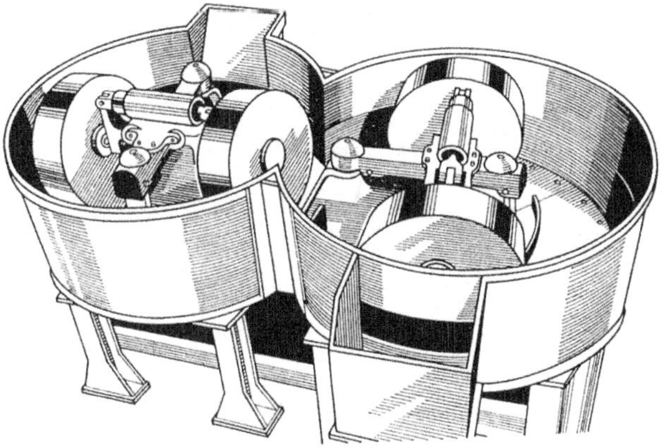

(*Courtesy August's Ltd.*)

FIG. 20. Continuous mixer.

on to move inside the mixing pan. As the rollers rotate any lumpy matters are crushed down while the ploughs churn the mix over and over ensuring a thorough mixing. Once mixing is completed, the material is discharged through a door at the bottom of the mixing pan. A design incorporates a spring-loaded muller arrangement which ensures maximum muller pressure with a minimum of muller inertia since the rollers are manufactured to light weight. The roller mills are designed to mix mould materials up to 4000 lb in weight, the usual sizes used being 2000 lb capacity. These have typical dimensions of 7 ft pan diameter to give a volume of 25 ft^3 with roller sizes 32 in. × 7 in. driven by a 40 h.p. motor. The mills can be fitted with automatic weighing hoppers and the doors can be operated pneumatically. A batch type mill involves charging the mill, emptying, recharging and so on. To eliminate time lost in this intermittent charging and discharging of the mixing pan, a continuous mixer has been designed which involves the use of two roller mills in series (Fig. 20). The two mills which are already mixing and contain the moist moulding material receive fresh sand, bond, etc., at a point such as 1 in Fig. 21. At position 2

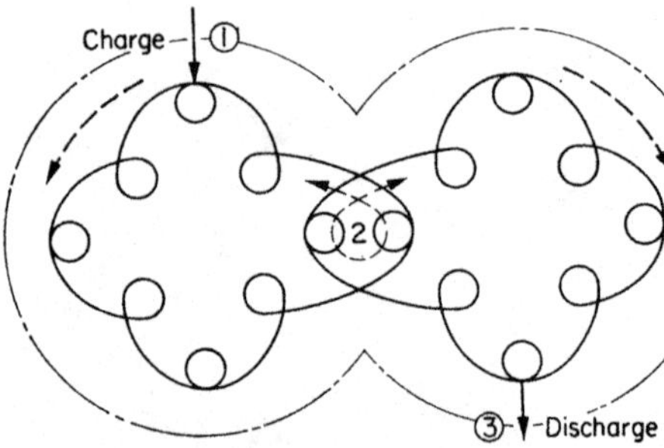

(*Courtesy August's Ltd.*)

FIG. 21. Mixing diagram of the continuous mixer.

the ploughs of the mills overlap and cause an interchange of moulding materials between the two mixers. The mixed material is discharged continuously over a dam (position 3) and is conveyed by belts to moulding machines. Charging of the continuous mixer with sand, bond, water, etc., can be made automatic and controlled to very precise limits.

Impeller Mill. An impeller mill is basically similar in construction to a roller mill with the difference that the rollers are replaced by a double- or three-blade paddle (Fig. 22). In this case the action is a simple mixing and any lumpy materials will not be crushed.

As the impeller rotates the mix climbs over the blades and hits two or more scrapers inside the mill. This gives a churning action and ensures uniform mixing. The impeller mills are normally employed for oil, silicate and air-set sands and are unsuitable for naturally bonded sands where a slight pressure from the rollers of a roller type mill is necessary to break down any lumpy clay matter and then distribute it evenly in the material being mixed.

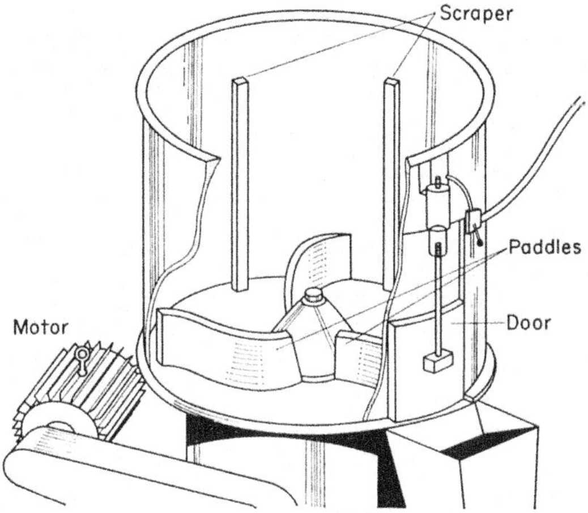

FIG. 22. Impeller type mill.

Method of Addition. For a mould material employing 100% natural sand, all that is necessary is to ensure thorough milling to break lumpy matters and to distribute the particles in the parent material. With synthetic sands it is preferable to moist the silica sand either with part or the whole of the water content to be added to the moulding material. This is to minimize any loss of bentonite and cereal in the form of dust when they are added to the mill. It is desirable that the clay or other additives are spread over the sand rather than charged on one point in the mixing pan. Spreading of dry additives is best achieved by keeping the mill running and introducing the dry additives in one point of the mill slowly. For large output mills, the cycle time between the preparation of the first and the following batch of sand can be reduced if overhead hoppers are fitted with the mill for the addition of bonds and other additives. The hoppers are mounted on overhead rails so that while the preceding batch of moulding material is being processed in the mill, the hopper can be loaded and brought over the mill for the second batch. Addition of bond and water must be carefully monitored.

A method of monitoring additives is by Pulse Jet Air Conveyor System whereby dry additives from hoppers are introduced into

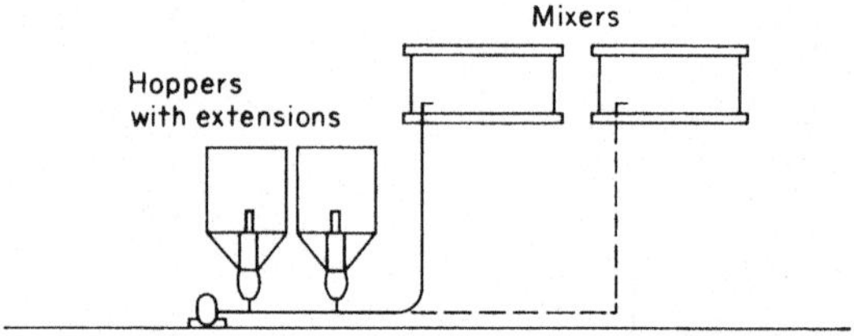

(*Courtesy August's Ltd.*)

FIG. 23. Pulse jet system.

the mills by air transport (Fig. 23) and can be used both for continuous and batch production lines. Figure 23 shows two hoppers but any number can be incorporated. The material, for example, bentonite is fed in the hopper (Fig. 24) and flows into the

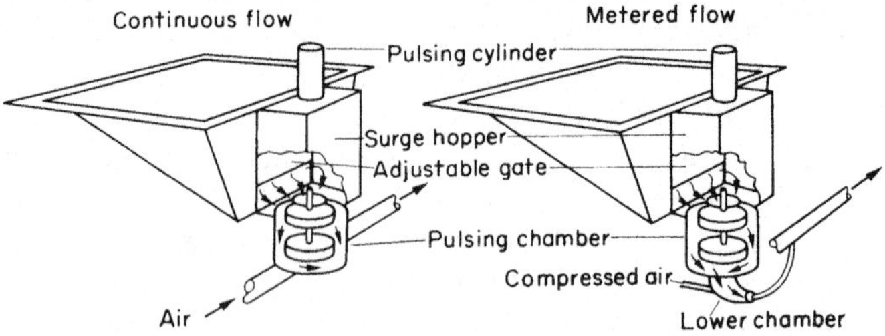

(*Courtesy August's Ltd.*)

FIG. 24. Pulsing chamber.

pulsing chamber through adjustable gates. The pulsing chamber incorporates a series of discs mounted on an extended air cylinder shaft. The material is metered by leading it from the pulsing chamber to a lower chamber where compressed air conveys the additive to the mill via air pipes. The dry additives are introduced at the bottom of the mill (Fig. 23) thus minimizing dust loss provided the sand in the mill is already moist.

Milling Time. Apart from lowering the rate of processing moulding materials from the mills, excessive milling time causes bond deterioration with all types of mould and core materials.

TABLE 16. VARIATION IN SAND PROPERTIES WITH INCREASING MILLING TIME

Milling time (min)	Moisture (%)	Green perm. No.	Green compressive strength (lb/in²)	Dry perm. No.	Dry compressive strength (lb/in²)
1	8·3	33	7·2	55	124
2	8·3	37	7·8	54	115
3	8·3	33	8·9	58	119
4	8·1	31	9·3	57	107
5	8·0	33	9·1	38	109
6	8·1	34	9·1	39	104
7	8·0	34	9·1	40	89
8	7·4	34	9·8	42	87
9	7·9	33	9·8	38	79
10	7·4	34	10·6	40	85
11	7·4	24	10·6	42	75
12	7·3	23	10·6	38	80

Typical data from a large roller mill preparing dry sand moulding material are shown in Table 16. The mix comprised a naturally bonded sand with an amount of silica sand and Wyoming bentonite. It is seen that, with increased milling time, there is a loss of moisture and a rapid increase in green strength.

The effect is to render the sand mix tough with reduced flowability of the mix during ramming of moulds. There is a loss of dry strength and the mould surfaces become friable upon drying. An optimum milling time for most mould and core materials is usually about 4 min.

Sand Plant Systems

For green-sand work moulds are prepared by hand ramming or are machine moulded. Usually, when castings are one off or very large as in dry sand moulding, moulds are rammed by hand. The latter types of moulds are also prepared by employing a sand slinger which gives a more uniform ramming than when hand rammers are used. Most large steel foundries have both jobbing and machine moulding sections.

Experience has shown that only a few inches thick facing of the mould material around the pattern need be carefully controlled from the point of view of green, dry and hot properties. Therefore, for economic reasons only the facing sand is prepared from clean new sand and the remainder of the volume in the moulding boxes is filled with backing sand. The backing sand must have an adequate degree of green and dry strength and permeability, but refractoriness and hot properties are not important. For this reason it is possible to employ used sands from the cast moulds as the basis material for backing sand preparation. The basic principle in preparing backing sands is to remove metal parts such as sprigs, core irons, etc., and lumpy sintered sand aggregates derived from the mould interfaces after casting. As described earlier, an amount of clay addition is made to compensate for the dead clay to the return sand while preparing the backing sand.

For large mechanized foundries, the whole sand system is so planned that it forms a complete cycle from the sand mill to the knockout area where the solidified castings are loosened free from the sand and moulding boxes. The general sequences in cycling sand through the foundry are shown in Fig. 25. The moulding machines are fed from two hoppers one containing the facing sand

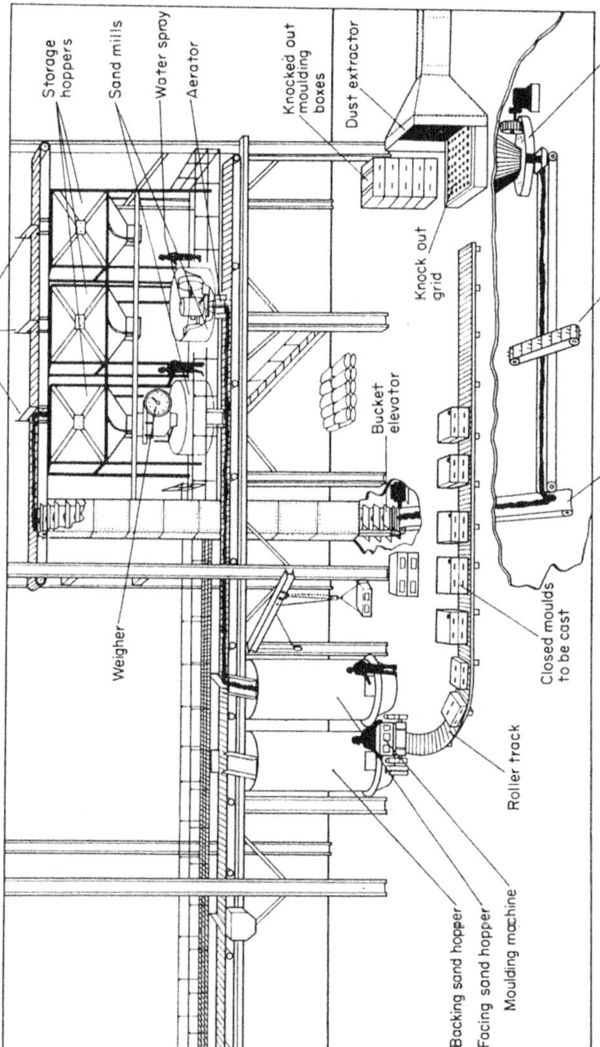

FIG. 25. Foundry sand system.

and the other the backing sand. Where a unit sand is used one single hopper per machine is employed. The rammed and closed moulds travel on roller tracks and after casting are conveyed to the knockout grid which is fitted with a dust extractor. The solid castings are shaken out to be dispatched to the fettling shop and the moulding boxes stacked and then returned to the moulding

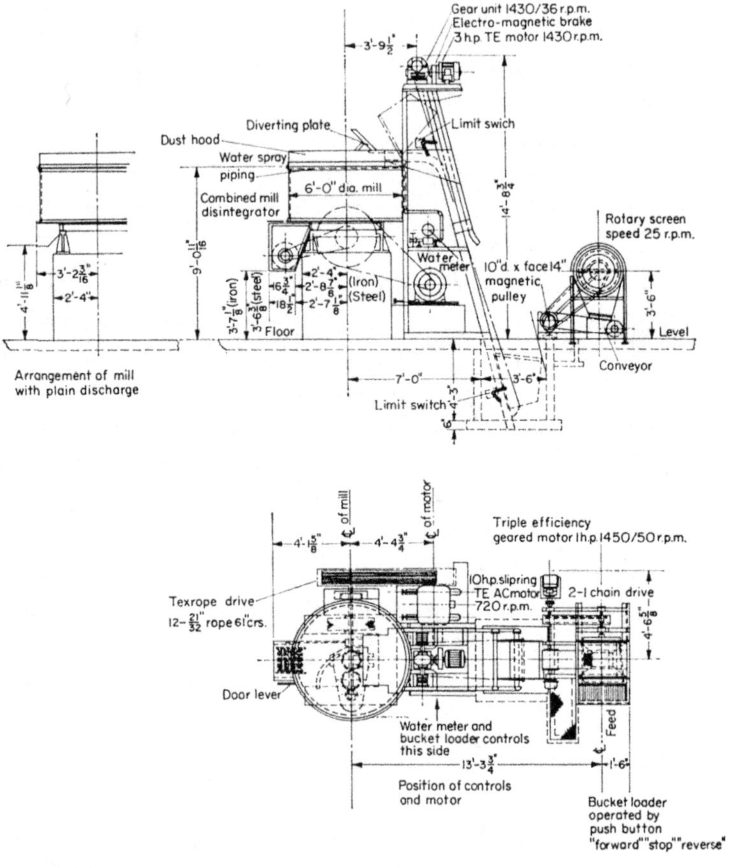

(*Courtesy Pneulec Ltd.*)

FIG. 26. Sand-conditioning plant.

machines. The sand falls onto a crusher where the lumpy materials are broken to free the silica grains and are then discharged onto a belt which passes under a magnetic separator. The magnetic separator removes iron particles and the return sand is conveyed by belts and bucket elevators to storage hoppers. From there it is discharged into the sand mill in amounts as required. For example, for unit sands 80% of the return sand and 20% of silica sand together with the necessary amount of bond, additives if any and water are added to the mill and the prepared sand is discharged, aerated and conveyed to the storage hopper over the moulding machine.

A complete sand plant can be quite elaborate in layout and the numbers of various parts such as mixers, hoppers, dust extractors, etc., will depend on the tonnage of sand consumed by the foundry. For small foundries a reconditioning plant as shown in Fig. 26 may be used. The return sand is fed manually into the rotary screen. This breaks lumpy matters and the loose sand grains fall on a conveyor, which discharges the material on a bucket loader after the tramp irons have been removed by a magnetic separator. The bucket loader discharges the sand into the mill which is fitted with water-spray pipes. Once processed the moulding material is discharged into a disintegrator which aerates the mix to improve its flowability. Any new silica sand, bond or additives may be added manually to the bucket loader. The water pipe is fitted around the inside of the mill so that the water spray is spread over the bulk of the sand in the mill.

CHAPTER 8

SHELL MOULD AND CORE MATERIALS

THE bonding medium in shell mould and core material is a resin. This is mixed with silica sand and the resulting material is dumped on a hot pattern plate or blown into a hot core box. The resin melts and as it flows an adhesive bond holds the quartz particles together. There are two broad types of resin, viz., (1) thermoplastic resin, (2) thermosetting resin. Thermoplastic resin such as natural wood rosin or synthetic resins such as polyvinyl chloride can be heated to melt and harden on cooling. The cooled solids can be reheated to remelt and this plasticity can be restored each time by repeated heating of the same cold resin. The thermosetting resins such as urea or phenol formaldehyde can be heated to flow and harden upon cooling. However, this cold resin cannot be remelted by a second cycle of heating, that is, unlike the thermoplasts the process is not reversible.

Resin Chemistry

Resins are polymers formed by recurring structural units called monomers. These can be natural resins such as are found in exudations of trees and they may not be true polymers. The true polymers are the synthetic resins and there are three main types, viz. thermoplasts, thermosets and rubbers.

Structure of Polymers. There are four basic structures of a polymer, viz.:

(a) linear structure;
(b) branched-chain structure;
(c) moderately cross-linked structure;
(d) highly cross-linked structure,

and are shown schematically in Fig. 27.

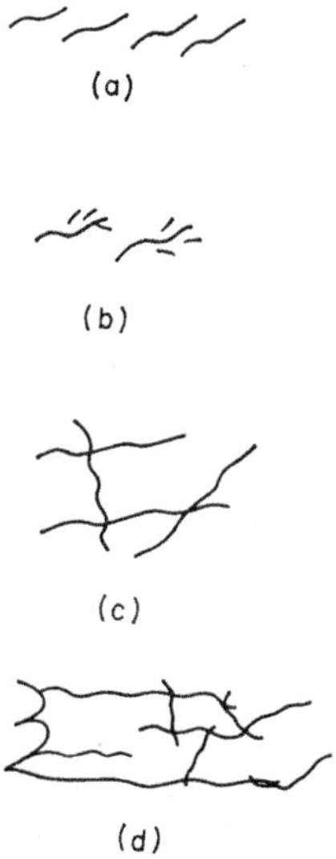

FIG. 27. Basic structures of polymers (schematic).

The linear structures (Fig. 27a) are mobile and have the ability to extend while presence of impurities causes branching (Fig. 27b). The cross-linked structure (Fig. 27c) also called loose network is more coherent than the linear or branched-chain structure. Cross-linking occurs by chemical reaction. The highly cross-linked structure (Fig. 27d) is characteristic of all thermosets and is also referred to as three-dimensional, mass or space polymers. The structure is coherent and the network has low mobility and extensibility.

The resins can be classified into two main groups such as:

(a) Phenolic resins.
(b) Amino resins.

Examples of phenolic resins are phenol formaldehyde, phenol furfural and resorcinol formaldehyde. Among the amino resins are urea formaldehyde and melamine formaldehyde.

Polymers are produced by either of the two following reactions, viz. addition polymerization and condensation polymerization or polycondensation.

Addition Polymerization. This reaction involves formation of repetitive structural units identical with the monomer, for example, formation of polyvinyl chloride from the monomer vinyl chloride,

$$nCH_2 = CHCl \rightarrow \left[\begin{array}{c} -CH_2CH- \\ \quad | \\ \quad Cl \end{array} \right]_n . \qquad (9)$$

Polycondensation. Unlike addition polymerization where all the reacting monomers have the same carbon to carbon bifunctional group, different groups are found in the monomers in polycondensation. Polycondensation is the operative mechanism in either the thermoplastic or thermosetting resins. The product of reaction between difunctional monomers is a thermoplastic resin while the thermosets are produced by reacting polyfunctional monomers. A highly cross-linked structure is produced if approximately equal molar quantities of tri- and difunctional monomers react.

Phenol Formaldehyde Resin. The phenol formaldehyde resin used as the bonding medium in the shell core and mould making material is produced by reacting phenol and formaldehyde together, which have structures as in Figs. 28 and 29. The points of double bond are the most reactive and the reaction is started by using an acid or an alkaline catalyst. The reaction proceeds in steps, the intermediate product being phenol monoalcohol. This

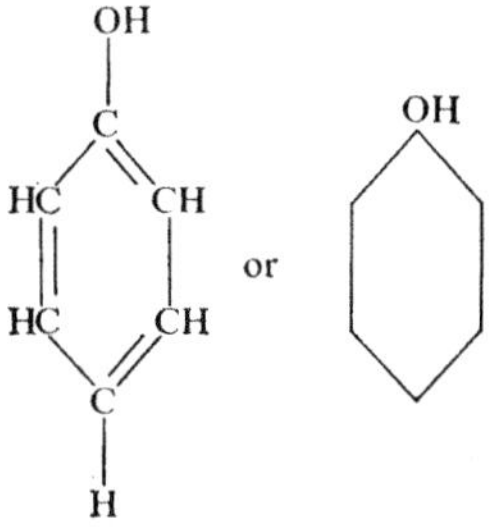

FIG. 28. Structure of phenol.

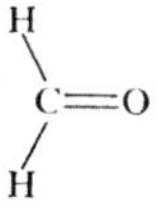

FIG. 29. Structure of formaldehyde.

reacts with excess phenol producing diphenyl methane and a molecule of water is split off. The phenol monoalcohol groups may also react together giving diphenyl ether and a molecule of water. Chain products are formed by further addition and condensation with the liberation of water and a mass molecule of phenol formaldehyde resin will form whose structure is shown in Fig. 30. The ratio of phenol to formaldehyde to form a mass molecule as shown in Fig. 30 is 1:1·5.

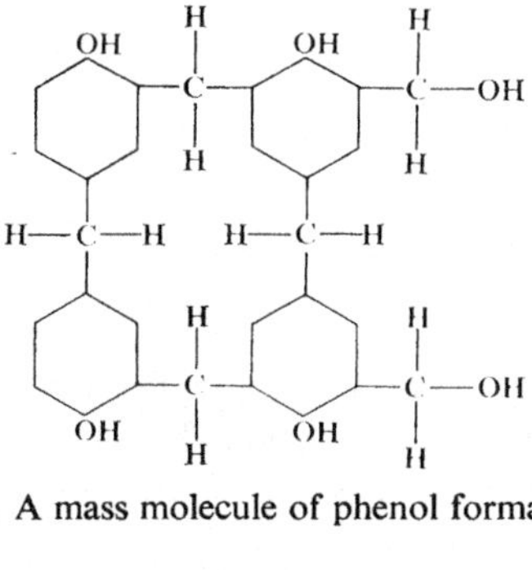

FIG. 30. A mass molecule of phenol formaldehyde.

One-stage Resin. Using the above ratio of phenol to formaldehyde and adding an alkaline catalyst, the one stage resin called a resole is produced. This is non-heat-hardenable.

Two-stage Resin. The resin used in the shell mould and core-making material is a two stage resin called a novalac. This is produced by using a lower ratio of phenol to formaldehyde group in the presence of an acid catalyst. The lower ratio prevents the formation of a highly cross-linked polymer. After the reaction between the two groups, the excess phenol and products of condensation are removed at a high temperature and a grindable powder is produced. This is the first-stage resin which is thermoplastic and is supplied to the foundry for preparing the mould or core material. The first-stage resin is mixed with silica sand, but in order to produce a highly cross-linked structure which is thermosetting, the necessary amount of formaldehyde in the form of hexamethylene tetramine is added to the ground first-stage resin. A mixture of the ground resin and hexamine as it is referred to is the bonding medium in shell-moulding mixes.

Sand Mixes

Apart from the bonding medium, viz. novalac and hexamine, a release agent such as silicone may be incorporated in the sand mix. The production of a shell mould is carried out by dumping the dry sand, novalac and hexamine onto a pattern plate which is kept at a temperature between 200 and 250°C. The bond melts forming a shell over the pattern plate, the thickness of the shell depending on how deep the required amount of heat has penetrated during the time the sand is in contact with the hot pattern. The mould or biscuit is ejected from the pattern plate by ejector pins. In order to facilitate this ejection, the pattern plate is sprayed with a release agent such as silicone. If this silicone is incorporated in the sand, production method can be simplified. The other commonly used additive is calcium stearate which allows the resin to spread more easily over the sand and also minimizes the clogging of the sand–resin mix in the mill allowing less load on the motor during sand preparation. Calcium stearate improves the flowability of the sand and gives increased tensile strength.

The resin content varies from 4% to 10% and the hexamine is about 10% by weight of the resin. The calcium stearate added is of the order of 1% by weight of the resin. Precoated sands require less resin and the resin content varies from 2% to 5%. A higher percentage of resin is used where excessively high strength is required. For example, the tensile strength is increased from about 160 to 400 lb/in² on increasing the resin from 2% to 4% for a silica sand having the major fractions in BS sieves 100, 120, 150 and 170. The resin content is as high as 10% where high finish and close tolerance is of prime importance. The resin content must, however, be kept low where casting design is such that gas porosity may occur readily.

Preparation of Resin-bonded Sands. Like clay bond or any other bonding material, it is imperative that the resin and additives are distributed evenly around the quartz grains. The simplest form of preparing the mould–core material is to mix them dry in the foundry.

Dry Mixing. This is done by using an impeller type mill and the mixing procedure is as follows:

(1) Charge sand in the mill.
(2) Add hexamine and mix until it is distributed thoroughly among the sand grains.
(3) Add novalac and mix until the resin is well distributed.

The impeller mills can be the vertical type as shown in Fig. 22 or rotating drums with paddles running horizontally. Roller type mills are not suitable as generation of heat must be avoided. Heating will cause partial melting of the novalac and the strength of the moulds and cores will deteriorate since the resin particles which have already melted will not melt again during investment of moulds or blowing of cores and bond strength is only developed as the resin flows upon the action of heat and cools subsequently.

One disadvantage of the dry mixing process is that milling time as long as half an hour is necessary to ensure thorough mixing. Moreover, during mixing, a cloud of dust forms which, apart from being uneconomical due to loss of resin, creates a toxic atmosphere.

Hot-coating Process. Although the dust can be suppressed by adding about 0·4% of a thin mineral oil which also prevents resin segregation, it is doubtful if very efficient distribution of bond around the sand grains is effected in the dry mixing process. Even if a perfectly uniform mix is achieved, segregation, because of the differences in density and particle size between the quartz grains and resin, inevitably occurs during repeated inversion of the dump box or blowing in core making. Segregation is at once apparent on the biscuits which show pockets having a mottled appearance indicating an impoverishment of resin there. To avoid this, the next stage of development was to produce a mould material comprising silica grains precoated with the bonding medium.

In the hot-coating process, the resin is allowed to melt partially and to flow around the sand grains forming an envelope of the polymer. It is necessary to use a heavy duty roller mill. The sand is first charged into the mill followed by the necessary quantity of novalac and hexamine. As milling proceeds, hot air is blown through the sand–resin mix at a temperature of the order of 150°C depending on the type of resin. The novalac and hexamine melt partially and flow around the sand grains. It is important that the temperature of the air is carefully controlled otherwise the resin will melt completely and, being thermosetting, bond development during the mould or core making stage will not occur.

The h.p. of the mill must be high because as the resin melts partially the mass in the mill becomes sticky and offers increasing resistance to the moving action of the rollers. At an appropriate stage when part of the resin has just melted to hold the remainder of the bonding medium around the sand grains a known quantity of water is sprayed on the sand–resin mix which lowers the temperature of the resin. Milling is continued until the lumps of the sand–resin mixes breakdown which are discharged through a screen and then to a cyclone via an air stream to be packed in bags or containers. The results obtained with hot, coated sands tend to be erratic possibly because of an amount of pre-curing during milling in hot air. It is also found that investment time of shell moulds is slightly longer than the dry sand–resin mixes.

Cold-coating Process. The difficulty of an amount of pre-curing as it happens during the hot-coating process can be overcome very easily by using the cold-coating process in which the resin dissolved in a volatile solvent is added to the sand together with hexamine and distributed around the quartz grains by milling. A roller type mill normally used in the foundry for clay-bonded moulding material is the most efficient. The mixing procedure is to charge the mill with clean silica sand and mix with hexamine until it is well distributed. The resin dissolved in the solvent is then added and milling continued. As mixing progresses the material becomes excessively sticky and the load on the mill increases as can be easily detected by connecting an ammeter to the motor. As the solvent evaporates, the sand–resin mix becomes loose and granular, which is discharged through a screen. It is desirable to have some means of extracting the solvent as it evaporates.

The cold-coating process is very easily adapted in a foundry as no special equipment is necessary. A limitation of the process, however, is that the investment time is prolonged during mould making, although this can be overcome by increasing the pattern plate temperature. There is also a possibility that in warm atmospheres the sand particles would cluster together giving rise to peel-back defect during dumping if the volatile solvent has not been completely removed from the mix.

Warm-coating Process. This process is exactly similar in principle as the cold-coating process with the difference that warm air is blown into the sand–resin mix to remove the volatile solvent. This results in decreased milling time and biscuits with high strengths are produced. The milling procedure is as follows:

(1) Charge sand into the mill. The sand temperature, although not critical, should not exceed 70°C.
(2) Add the necessary quantity of hexamine together with additives, for example, the release agent. Mill for half a minute for uniform distribution.
(3) Add resin and mill for 2 min.
(4) Commence blowing air which is at a temperature between 200 and 230°C into the sand–resin mix.
(5) Continue milling. A stage will be reached when the load on the motor will be a maximum. This will be followed by the particulate matters breaking loose.
(6) Discharge through a $\frac{1}{16}$ in. screen and cool, when it is ready for using in the foundry.

The mixing procedure can be completely automated and the whole operation is of the order of 10 min.

The Nature of Bond

The principle of providing the bonding force between two particles coated with resin is to heat to a temperature where the dry resin melts and flows which, upon cooling, holds the particles together. It is shown that the bonding force is adhesive in nature and the work of adhesion W_{AD} (dynes/cm) can be calculated from the equation

$$W_{AD} = \gamma (1 + \cos \theta), \tag{10}$$

where γ = surface tension of the liquid resin at the temperature, and θ = the contact angle in degrees between the solid and the liquid.

It is shown that zircon gives a higher value of W_{AD} with phenol formaldehyde than silica and the corresponding tensile strengths are higher. The nature of the adhesive bond requires that sand surfaces should be clean to develop optimum bonding and strength deterioration occurs in the presence of impurities or additives such as clay, iron oxide, silica flour, etc. The strength values of shell moulds increase with increasing quantity of resin; however, the amount of gas evolved is also higher. The tensile or transverse

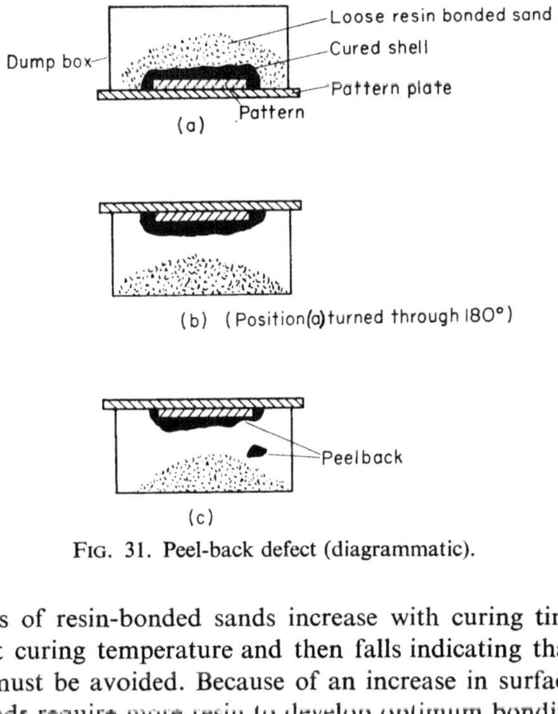

FIG. 31. Peel-back defect (diagrammatic).

strengths of resin-bonded sands increase with curing time at a constant curing temperature and then falls indicating that over-curing must be avoided. Because of an increase in surface area, finer sands require more resin to develop optimum bonding than coarse sands.

Peel-back

The ideal resin bond is that which will form a rigid shell which will remain stuck to the pattern plate. The peel-back defect is that where part of the shell drops off from the biscuit. This is shown diagrammatically in Fig. 31. In (a) the resin has cured on the hot pattern plate forming a shell, while at (b) the dump box has been inverted leaving the cured shell adhering to the pattern plate and the uncured loose sands have fallen into the dump box giving a biscuit without the defect. In (c) peel-back has occurred where part of the shell has detached from the parent biscuit.

Peel-back defect has been explained in terms of the flow characteristics of the resin employed for bonding. In the case of phenol formaldehyde resin, the novalac melts in a vapour of hexamine which sublimes. As the novalac melts it flows until it

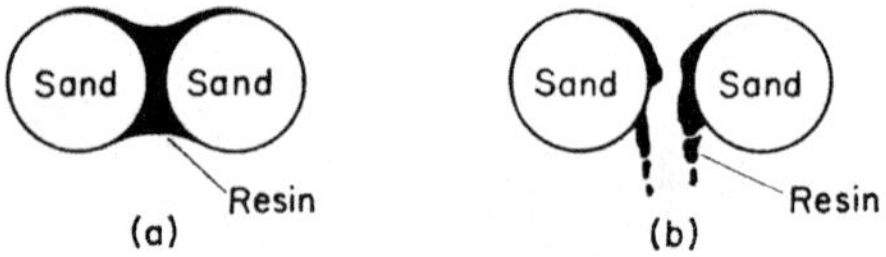

FIG. 32. Long- and short-flow resins around sand grains (diagrammatic).

gels by reacting with hexamine. The distance of flow determines whether the resin is a long-, medium- or a short-flow resin. A short-flow resin forms a more coherent meniscus (Fig. 32a) giving a better bond than a long-flow resin where part of the resin drains away (Fig. 32b) causing an impoverishment of resin at the points of contact between the sand grains, and peel-back defect occurs in the latter case.

Bibliography

B. GOLDING, *Polymer and Resins*, Van Nostrand Co. Inc., Princeton, N.J., 1959.

R. W. MARTIN, *The Chemistry of Phenolic Resins*, John Wiley & Sons, Inc., New York, 1956.

N. J. L. MEGSON, *Phenolic Resin Chemistry*, Academic Press Inc., New York, 1958.

J. K. SPRINKLE and H. F. TAYLOR, *Trans. AFS*, **65**, 300 (1957).

P. H. R. B. LEMON, *Foundry Trade Jl.* **101**, 113 (1956).

CHAPTER 9

FURANE AND OTHER BINDERS

THE difficulty with the thermosetting resin is that heated pattern plates and curing ovens are necessary. Therefore large castings have not been a practical proposition and shell mould and core making have been confined mainly to small castings. The resin binders which can be hardened by simply standing in air have found very useful application in the production of large moulds and cores in the steel foundry. Like the carbon dioxide process, the core boxes are rammed and hardened *in situ*. This means that stripping of cores and their consequent drying in stoves are eliminated resulting in a saving of labour and fuel cost. Strengths of self-curing or carbon dioxide hardened materials are comparable to conventional core sands. However, these non-stoving binders are relatively expensive.

Cold-curing Process

Cores which are rammed and hardened *in situ* by standing in air have the core material composed of silica sand and a group of resins which form highly cross-linked polymers upon the action of acids. The resin binders are made by mixing various proportions of urea, formaldehyde and furfuryl alcohol ($C_4H_3O{-}CH_2OH$). A mixture such as this is used for reasons of economics and also because furfuryl alcohol alone gives a relatively weak bond. A strong acid such as phosphoric acid is used to increase the curing speed of the rammed moulds or cores.

Experience has shown that the presence of urea gives rise to pinhole porosity in steel castings. Therefore, for steel foundry application the urea is omitted from the compound binder and the cold-curing or no-bake binder for steel foundry use is essentially a partially polymerized furfuryl alcohol mixed with 15% of paraformaldehyde.

The acid is referred to as catalyst and in some cases toluene sulphonic acid is also used. The popular catalyst or hardener, as it is often called by proprietary suppliers, is phosphoric acid

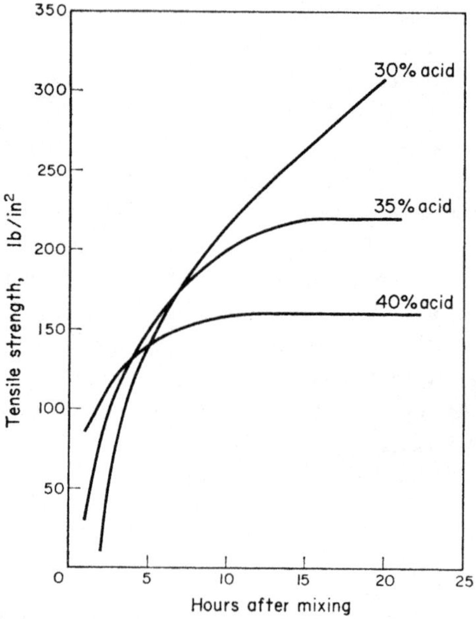

(*Courtesy F. & M. Supplies Ltd.*)

FIG. 33. Effect of various amounts of an acid catalyst on a furane bonded sand.

(sp.gr 1·7) and a typical sand mix is clean silica sand, 2% binder and a catalyst addition of between 30 and 40% by weight of the binder. The setting time of the cores is governed by the amount of catalyst and the sand temperature. As the catalyst is increased, higher strengths are achieved at the early stages, but the ultimate strengths are lower with a higher amount of catalyst addition (Fig. 33).

Whether a faster curing rate and hence increased production or a higher ultimate strength is required, will be decided by the requirements of the individual foundry. However, a stripping time of cores can be successfully established by noting the tensile strength against standing time in the laboratory. A tensile strength of the order of 60 lb/in² is a good guide, and using this a graph such as in Fig. 34 can be used to estimate the stripping time in the

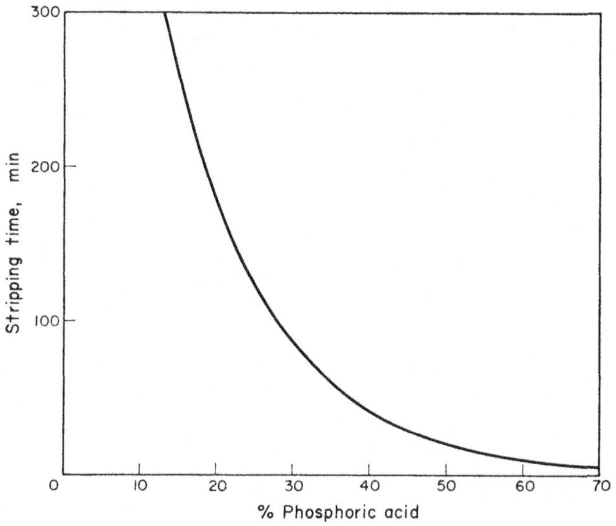

(*Courtesy Harborough Construction Company Ltd.*)

FIG. 34. Stripping time with increasing amounts of catalyst.

core shop. The rate of curing and hence stripping time will also be controlled by the sand temperature as it would vary between seasons, and this is shown in Fig. 35 for a particular sand with 1·5% proprietary binder and 40%, 1·70 sp.gr. phosphoric acid. A knowledge of stripping time is important because the core boxes are released for a second core and the first core can stand in air for complete hardening.

The acid is corrosive and care must be taken in handling it.

When wooden core boxes are used, they should be varnished with an epoxy resin based material and shellac should not be used. The binder can usually be stored for 3 months in a cool place. The accelerator can be kept indefinitely. In the winter, however, the phosphoric acid would freeze but can be easily liquefied by gentle warming. The catalyst and binder must never be mixed together as the reaction is exothermic and violent.

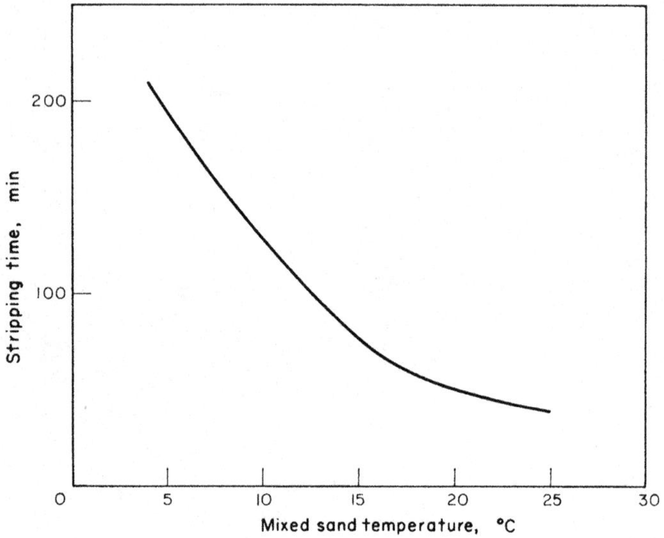

(Courtesy Harborough Construction Company Ltd.)

FIG. 35. Stripping time with increasing sand temperature.

Sand Mixing. The core material should be mixed with an impeller type mill and the mixing procedure is as follows:

(1) Add clean silica sand to the mill.
(2) Add catalyst and mix until evenly distributed.
(3) Add binder and mix and discharge for mould or core making.

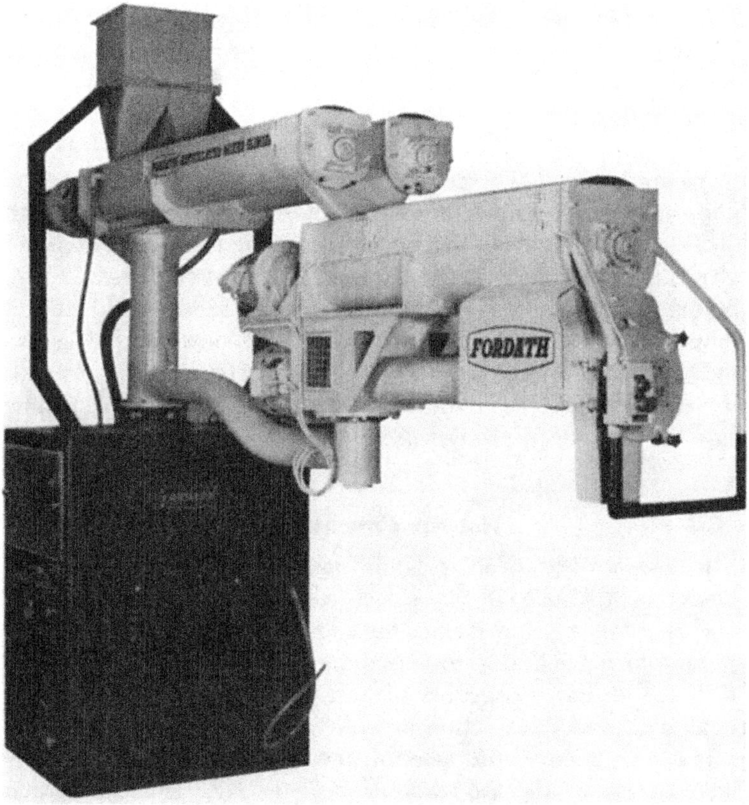

(*Courtesy the Fordath Engineering Company Ltd.*)

FIG. 36. Continuous mixer slinger.

Curing of the sand mix begins in the mill as soon as the binder and the catalyst come in contact, and the bench life of these mixes is very poor. The limitation imposed by a poor bench life of this chemically hardened core material can be overcome by using the continuous mixer slinger (Fig. 36). The sand is introduced through the sand hopper into the mixing trough fitted with helical blades. Both the binder and catalyst are fed into the sand by pumps and the mixed core material can be discharged directly into core boxes by using a slinger head. The discharge end can be moved in an arc and a whole range of core boxes can be rammed in a short period of time.

Nature of Sand. The condition of the silica sand employed is important. For example, sands of marine origin contaminated with alkaline salts and shells and those containing iron oxide and clay give variable results with the catalyst and, in extreme cases, curing becomes impossible. It is desirable that the sand should have a minimum of 95% SiO_2. As already discussed warm sands decrease the bench life. Very cold sand in the winter delays curing time and, in extreme cases, the inside of the core does not harden. In this case the amount of catalyst needs increasing.

Hot-box Core Material

In the hot-box process sands bonded with similar types of binder as those used in the cold-curing process are employed, but a weak catalyst is added since the core material is blown into core boxes which are heated to temperatures in the range 200–250°C. Furfuryl alcohol is capable of polymerizing in the presence of weak acid under the action of heat from the core box. Like the cold-curing binder, the alcohol and acid form a highly cross-linked polymer and the reaction is exothermic. This exothermic reaction is useful as the heat liberated facilitates curing right through the centre of the core.

The furane resins are expensive, and binders developed from sugar and urea formaldehyde have been developed in France. The binders contain between 30 and 50% water. The viscosity of the binders must be low to ensure complete dispersion among the sand grains. As in the cold-curing resin bonded mixes, the silica sands should have a high SiO_2 content and be free from shells, alkalis, clays and mineral oxides. Zircon and olivine have been used successfully.

Since the mixed sand is used in core-blowing machines, it must have good flowability. The green strengths of the mixes are low and the amount of water content in the core material is critical. Too low a moisture content results in friable cores, while excess water increases baking time. Since the core material is stored in hoppers, a good bench life is important and the mixes are formulated to give a bench life of 24 hr. The bench life is a function of the amount of acid catalyst used, and the sugar-based binders have very long bench life unlike the urea formaldehyde. The quickest curing time is obtained with a mixture of urea-formaldehyde and sugar-based binders. However, because of the possibility of pin-hole porosity, urea formaldehyde cannot be used in steel foundries.

At present there is a world-wide shortage of furfuryl alcohol and, in Great Britain, proprietary binders based on phenol are available for producing cores for the steel foundry by the hot-box process. A proprietary phenol formaldehyde based resin requires a 2% binder addition, although this needs increasing if the sand is excessively fine. The resin is capable of hardening by itself but the rate of curing can be accelerated by introducing 10–25% of toluene sulphonic acid catalyst by weight of the binder added. Typical tensile strength data for holding time in core boxes of 10, 20, 30 and 60 sec are 25, 100, 150 and 360 lb/in^2 respectively. A comparable tensile value for a 2% sugar-based binder is of the order of 300 lb/in^2.

The milling procedure of hot-box core material is similar to cold-curing resin bonded sands, and similar precautions in handling the acid are necessary.

Sodium Silicate–Carbon Dioxide Process. Probably the first serious thoughts on the use of mould and core materials which could be rammed and then allowed to develop optimum strengths *in situ* were directed on the use and development of sodium silicate bonded sands hardened chemically by passing carbon dioxide gas in the rammed moulds and cores. The process is well established in steel foundries and a very detailed account of the mould and core making principles has been given in an earlier book issued in this library.*

The basis material used is clean, cold silica sand to which is added an amount of sodium silicate, $Na_2O.SiO_2.xH_2O$. The sodium silicates are specified in terms of their SiO_2/Na_2O ratios, and a silicate with SiO_2/Na_2O ratio of 2 is popularly used in British steel foundries. Silicate-bonded sands have little green strength and a major limitation of the process is its very poor breakdown performance.

It has been demonstrated, however, that good breakdown of cores after casting is possible if a mineral additive such as rutile or alumina is used in conjunction with an organic binder viz. sugar or starch. In British foundries, it is customary to use a proprietary binder and add an amount of ball clay. The clay is used chiefly to impart an amount of green strength to the mix where it is thought necessary to gas the moulds after stripping the pattern. An amount of clay assists in the collapsibility of the material during knockout probably because of the presence of alumina. Silicate-bonded sands have very good bench life when compared with furane binders and can be stored as long as a week in sealed containers. If exposed to the atmosphere, the material becomes hard because of dehydration.

A typical mix used in the steel foundry is as in Table 17. The water is added as both the clay and the breakdown powder are

TABLE 17. A SODIUM SILICATE BONDED SAND MIX

	%
Silica sand	93·5
Water	0·5
Sodium silicate (SiO_2/Na_2O = 2)	4·0
Ball clay	1·0
Proprietary breakdown powder	1·0

* A. D. Sarkar, *Foundry Core and Mould Making by the Carbon Dioxide Process*, Pergamon Press, Oxford, 1964.

hygroscopic and tend to absorb part of the moisture from the aqueous silicate necessitating an excessive amount of sodium silicate addition. This, apart from being uneconomical, decreases the breakdown efficiency of the sand. Only a minimum amount of water should be added, otherwise the moulds and cores will harden only with difficulty when gassed with carbon dioxide.

All additions and the silica sand should be weighed accurately, and the milling procedure should be as follows:

(1) Add sand in the mill. Impeller type mills are preferred, but roller type mills have been used successfully.
(2) Add water and mix for half a minute.
(3) Add ball clay and breakdown powder and distribute well by milling. The time required is usually 1 min.
(4) Add sodium silicate and mill for 3 min.

The silicate-bonded sands like the furane binders leave sticky residues in the mill. It is well to make a practice of cleaning the mill with water at the end of a shift. It is preferable to use separate mills for clay- or oil-bonded sands and silicate- or furane-bonded mould and core materials.

The Ferrosilicon Process. A modification of the sodium silicate–CO_2 process has been to incorporate additives in the sodium silicate–silica sand mix which would cause self-hardening in the rammed moulds and cores. That is, gassing of moulds and cores will be unnecessary.

In the ferrosilicon process, the additive employed is very fine ferrosilicon and hardening appears to occur by an exothermic reaction causing gelation of the sodium silicate. In 1963 Takio Nishiyama presented a paper* where some examples of production iron and steel castings were given. It was reported that a sand mix containing 6·5% sodium silicate of SiO_2/Na_2O ratio 2·0 and

* Takio Nishiyama, exothermic self-hardening moulds made by the Nishiyama process, *Proceedings 30th International Foundry Congress, Praha*, 1–6 September, 1963, pp. 163–76.

2% 75/25 ferrosilicon was used, the ferrosilicon being BS – 300 mesh size.

The present author's experience shows, however, that, although the optimum particle size of the ferrosilicon is BS – 300 mesh, such a large quantity of sodium silicate is unduly high for the normal range of sands used in British steel foundries and the material ferrosilicon is expensive. A 3% sodium silicate of SiO_2/Na_2O ratio 2 is found to produce good workable mixes provided the dry sand is milled with 0·5% water previously. Table 18 shows that reasonable strength is achieved with 1% 75/25 ferrosilicon addition although higher strengths are experienced when the amount of additive is increased.

TABLE 18. STRENGTH–STANDING TIME RELATIONSHIP FOR FeSi AND CaSi BONDED SANDS

Standing time (hr)	Compressive strength (lb/in²)				
	FeSi %			Ca–Si %	
	0·5	1·0	1·5	0·5	1·0
0	–	–	–	2·0	–
½	–	–	90	16·5	–
1	12	85	195	48·0	35
1½	17	152	250	54·0	–
2	–	–	275	72·0	51
2½	–	–	–	–	–
3	–	–	–	93·0	90
3½	145	–	–	–	–
4	–	345	–	120	90
5	210	–	–	–	91
6	–	–	–	279	–

However, even at 1% ferrosilicon addition, the mould material is more expensive than the CO_2 process assuming a gas consumption of 3 wt. % of the total sand. The bench life of the mix is poor. However, ferrosilicon bonded sands appear to give somewhat superior collapsibility to carbon dioxide hardened cores. Some limited experience with small jobs has shown that castings free from blowholes and scabs can be produced, although some surface blemishes are apparent, possibly a surface deoxidation effect.

It is probable that ferrosilicon can be replaced by calcium silicide or some other exothermic reaction forming additives. Table 18 shows, however, that, for similar amounts of addition, calcium silicide of comparable grain size gives inferior properties to ferrosilicon.

Other modifications of carbon dioxide hardened mould materials such as the lime–silicate or the liquid sand process do not appear to have found much application in British steel foundries. In the former, powdered lime, sand and sodium silicate are mixed together and hardening is believed to occur through the core because of an exothermic reaction by lime.

In the liquid sand process, the silica sand is foamed by mixing with a quantity of water and a trace amount of Teepol. A low viscosity high ratio (greater than 2) sodium silicate is then mixed with the sand. Once distributed, very fine dicalcium silicate, the hardening agent, is added and, after mixing, the slurry is poured over the pattern which is withdrawn when the mould material has set. A cement mixer is found very suitable for preparing the liquid sand. First a Teepol–water stock solution containing 100 parts of water and 8 parts of Teepol by weight is made and a composition of liquid sand is given in Table 19.

TABLE 19. A LIQUID SAND MIX

100	parts by weight of sand
2	parts by weight of Teepol–water stock solution
5	parts by weight of high ratio sodium silicate
4	parts by weight of dicalcium silicate

An addition of 4% dicalcium silicate appears to be optimum, although this would vary according to the grade of the hardening agent used. During mixing it is necessary to observe the sand and Teepol mix. As soon as it "melts", the sodium silicate should be added and distributed. The dicalcium silicate needs to be dispersed quickly followed by casting of the mould by the mix, otherwise the sand will harden and become powdery in the mixer. It is important that the moulds are allowed to stand in air to dehydrate; the standing time may be of the order of 24 hr. Freshly cast moulds cause dangerous bubbling of moulds. The cores produced by the process become friable if the bubbles produced by the Teepol solution are excessively large. The size of the bubbles can be controlled by adding a trace amount of coconut diethanolamide, a lather booster.

Ethyl Silicate Bond. The ethyl silicate or phosphate binders provide the bonding agents in the field of investment castings. The lost wax process involves the production of a ceramic shell ranging in thickness up to ¼ in. The basis grains used are of fine particle size as surface finish of cast components is an important consideration. The binder employed to hold the particulate matters is usually a water–base silica sol. The pattern is made of wax and a shell is built by alternate dipping and air-drying in slurries of controlled viscosity. Only the facing is made with fine (BS—300 mesh size) particles and is built after about two dips. The subsequent coats are from coarser particles. Once a shell forms the wax is removed by melting. The precision castings such as turbine blades are cast in groups from a common sprue, and after removal of the wax, the whole assembly is fired at a temperature of the order of 900°C and the moulds are cast while hot. The other application of ceramic mould material is where a mould is produced in the conventional way by using a metal pattern rather than the expendable wax pattern. In this case one slurry is produced and poured over the pattern, the mould being stripped after the material has set.

Components manufactured by ceramic-mould process are produced to close dimensional tolerance, and investment casting gives the notion of being a field on its own rather than being a part of a steel foundry in general. It is beyond the scope of this book to consider the material in detail here, but the principles of ethyl silicate bonding in so far as it covers mould production by using a metal non-expendable pattern are outlined here. It should also be mentioned that some thoughts are being given to the use of ethyl silicate bonded alumina type materials as facing materials, for heavy steel castings.

The Basis Material. An important consideration in choosing the basis material is that it must be stable at high temperature and the non-siliceous materials such as sillimanite, mullite, alumina and zircon possessing low expansion characteristics are preferred to silica while preparing the investment mould materials for manufacturing castings of high dimensional accuracy. High thermal expansion of silica is known to cause cracks in the moulds during heating.

A mixture of fine and coarse particles gives the ideal slurry. Too small a particle size gives high hot strengths and hence cracks in the casting during cooling. On the other hand, coarse particles tend to sink, and a uniform dispersion of the particulate matters in the slurry does not occur. Even a mixture of coarse and fine particles gives average grain size well below, for example, the clay-bonded sands. This, together with the nature of the bond, creates a high hot-strength material. The materials, however, are known to have low thermal conductivity resulting in a relatively low thermal gradient in the cooling castings and hence low cooling stress. This factor minimizes hot tear propensity.

Ethyl Silicate

Ethyl silicate hydrolyses, for example, upon acidification with HCl in the presence of water. The product of hydrolysis is that silica is deposited and an amount of alcohol is given off. The silica part of the reaction is adhesive in nature and provides the bonding force between the filler grains.

Hydrolysis of an ethyl silicate solution as given in Table 20 is affected by allowing it to stand for two hours with frequent stirring.

TABLE 20. ETHYL SILICATE SOLUTION

Ethyl silicate	37·6% by volume
Ethyl alcohol	59·8% by volume
3% HCl	2·6% by volume

The hydrolysed solution is mixed with the aggregates of, for example, sillimanite, and the slurry is poured over the pattern. The moulds are stripped once gelation sets in. The gelling time is controlled by the acid concentration of the solution and often about 0·3% MgO in very fine form is added to the basis material to neutralize the effect of the acid and give a more controllable gelling time. The gelling time can also be controlled by adding NaOH, NH_4OH, piperidine, etc.

Experience is necessary to assess at what stage of gelation the moulds should be stripped from the pattern. Casting of moulds should be done under vacuum or alternatively the slurry should be poured on patterns resting on a vibrating table to remove trapped air. After stripping, the alcohol is burnt off by simply igniting the mould in air followed by firing at a temperature of 1000°C. This firing is necessary to dehydrate the mould completely. An effect of this high temperature is also to produce a strong sintered bond between the refractory particles.

Bibliography

P. JASSON, *Brit. Found.* **54**, (7), 309 (1961).
P. LEMON, *Ibid.* **54** (12), 542 (1961).
R. S. BUSHNELL and E. PARKES, *Ibid.* **55** (8), 325 (1962).
W. G. LAWRENCE, *Trans. AFS*, **68**, 455 (1960).

CHAPTER 10

HEAT EXTRACTION BY MOULDS

THE mode of solidification of a casting, *ceteris paribus*, is governed by the ability of the mould to extract heat from the casting. If the rate of heat extraction is rapid, the metal interface temperature will be held at a relatively low temperature resulting in a steep temperature gradient in the freezing metal and the casting will solidify in a progressive manner towards the heat centre, a condition necessary for maximum soundness in a casting.

Mathematical Treatment. The basic equations for heat extraction by moulds are those of unsteady heat flow and it is known that for an isotropic homogeneous medium of dimension $2dx$, $2dy$ and $2dz$ through which heat is flowing,

$$\frac{\partial \theta}{\partial t} = \alpha \left(\frac{\partial^2 \theta}{\partial x^2} + \frac{\partial^2 \theta}{\partial y^2} + \frac{\partial^2 \theta}{\partial z^2} \right), \tag{11}$$

where θ = temperature of the medium; t = time; α = thermal diffusivity of the system; and $\alpha = K/\rho c$; and for the solid medium K = thermal conductivity; ρ = density; and c = specific heat.

When it is assumed that heat is flowing through one direction only, for example along x, eqn. (11) becomes

$$\frac{\partial \theta}{\partial t} = \alpha \frac{\partial^2 \theta}{\partial x^2}. \tag{12}$$

Solution of the basic equation for unsteady heat flow shows that

$$\theta_m = \theta_0 + (\theta_i - \theta_0) \left[1 - \operatorname{erf} \frac{x}{2\sqrt{(\alpha t)}} \right], \tag{13}$$

where θ_m is the temperature at a time t for a point at a radial distance x from the interface whose temperature is raised instantaneously to θ_i and whose ambient temperature was θ_0. The term erf $x/[2\sqrt{(\alpha t)}]$ is known as the error function.

Equation (13) is valid provided it is assumed that the interface temperature remains constant with time and the thermal diffusivity of the mould material does not alter greatly with variation in temperature. It is also assumed that the errors introduced by the non-uniformity and the finite size of the mould are not large.

By using eqn. (13), it can be shown that the total heat absorbed by a mould surface of unit area in time t is

$$Q = \xi\sqrt{t}, \tag{14}$$

where $\xi = 1{\cdot}128\text{b}\,(\theta_i - \theta_0)$ and $\text{b} = \sqrt{(K\rho c)}$, the heat diffusivity of the material. In spite of the assumptions made, eqn. (14) can be applied quite successfully to sand moulds. However, calculation of heat extracted by a mould involves a knowledge of density, specific heat and thermal conductivity of the mould material together with data on interface temperature. None of the parameters can be obtained simply, but eqn. (14) can be utilized in a different way.

Thus, if for two different materials the subscripts 1 and 2 are used and the quantity of heat absorbed by the same steel cast at similar temperatures is Q_c,

$$Q_c = \xi_1\sqrt{t_1} = \xi_2\sqrt{t_2} \tag{15}$$

or

$$\frac{\xi_1}{\xi_2} = \sqrt{\left(\frac{t_2}{t_1}\right)} \tag{16}$$

If the time t for each material is taken as that for the solidification of a standard casting, then the relative chilling power of a mould material can be measured in the following way. A standard casting, for example, a 6 in. diameter × 6 in. high cylinder is chosen. A mould is made using a reference material, for example, 5% sodium bentonite bonded dry sand. By means of a thermocouple at the centre of the mould the complete solidification time of the casting t_{ref} is noted. A similar experiment is carried out using the material whose chilling power is being investigated by using a similar steel cast at the same temperature and ramming the mould to the same bulk density as the reference mould. The solidification time t_{mat} of the casting produced in the material is noted, and the relative chilling power of the material is expressed as

$$\text{Relative chilling power} = \sqrt{\left(\frac{t_{ref}}{t_{mat}}\right)}. \qquad (17)$$

As the ratio in eqn. (17) increases, the chilling power of the material increases.

A knowledge of the heat extracting capacity of various materials is important since directional solidification in a casting with unfavourable design can be improved by using mould materials of differing heat extracting capacities in the same mould. It is doubtful, however, if this is done very methodically in many steel foundries although the use of external and internal metal chills to accelerate solidification in parts of a casting is very common. Moreover, it is not clear if the usual silica or non-siliceous-based materials bring about any significant change in the solidification times in large and heavy steel castings. However, experience shows that zircon and chromite-base materials provide a cleaner strip for castings than the silica base material. For low casting temperature some surface lapping effect is also noted with the former two non-siliceous materials. Both these factors indicate a higher chilling power of zircon and chromite than silica.

Bibliography

R. W. RUDDLE, *The Solidification of Castings*, the Institute of Metals Monograph and Report Series No. 7, 1957, 2nd edition.

C. W. BRIGGS, *Trans. AFS*, **68**, 157 (1960).

CHAPTER 11

SAND RECLAMATION

UPON and after the entry of liquid steel, the mould interface heats up, and the temperature within the mould rises, parts removed from the interface having progressively lower temperatures. Time–temperature distribution for a heavy 10 ton casting is shown in Fig. 37. The side walls heat up to higher temperatures and to a greater depth than the cope of the mould which receives radiant heat only during the time steel enters the mould. According to the temperature distribution the side walls of a mould can be divided into four zones (Fig. 38) viz.:

(1) *Zone 1.* This is the zone adjacent to the mould interface possibly not more than $\frac{1}{2}$ in. thick, which has attained a temperature exceeding 1200°C.

(2) *Zone 2.* This is the zone where the temperature attained by the moulding mix is probably in the range 500/600 to 1200° C, and may be of the order of 3 in. thick depending on the casting section.

(3) *Zone 3.* This zone begins where zone 2 has ended and the temperature is such that all the adsorbed water has been removed.

(4) *Zone 4.* This is the part whose outer boundary is defined by the moulding box and its moisture content is relatively unaffected.

Sand from the last zone then can be re-used for moulding, whereas the sand–clay agglomerate from zone 3 need only be hydrated to the optimum level and new moulds can be rammed. The clay and quartz in zones 1 and 2, however, have undergone changes. In the low temperature range, the clay has lost its constitutional water and cannot be rehydrated so easily. Further, at higher temperatures the clay has sintered and melted and the quartz particles have undergone allotropic and volumetric changes.

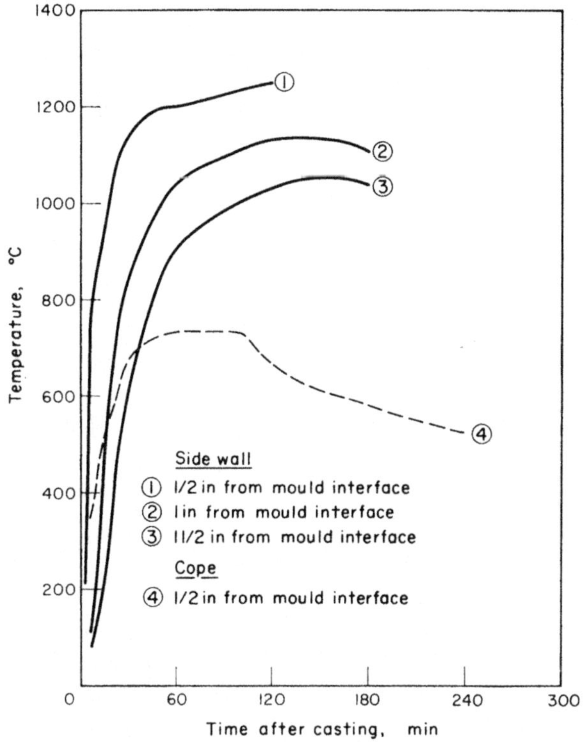

FIG. 37. Temperature–time curves at various distances from the mould interface for a 10 ton steel casting.

Reconditioning

For economical reasons mould and core materials, whether in the form of silica or naturally bonded sands, should be re-used at least partly. The physical condition of the aggregates falling through the knockout grid is broadly usable clay-bonded quartz grains and those needing fresh clay–water bond together with metal parts such as core irons, sprigs, etc. Further, part of the sand–clay aggregate is in lumpy form because of the sintering of the clay bond at high temperatures; the volumetric expansion on heating and contraction on cooling may cause fracture of the quartz grains resulting in an amount of fines in the material. The dehydrated clay is also a source of fines.

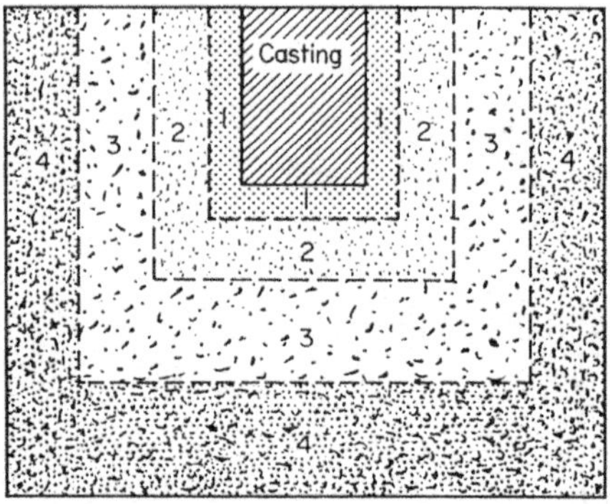

FIG. 38. Zones behind a metal–mould interface (schematic).

In the reconditioning process, the return sand itself is re-used after removing the lumpy matter, the fines and the tramp metallic elements. This is done by first crushing the sand in a heavy roller mill or tumbling in a rotary ball or chain-type mill. The sand is discharged to pass under a magnetic pulley to separate the irons and then screened to remove oversize lumpy matters which have not been crushed in the roller or rotary mills. The fines can be removed by passing a stream of air at a suitable stage, e.g. while the return sand is being crushed in the mill. The reconditioned sand can be remilled to process a second batch of moulding material provided an approximation is made regarding the amount of clay lost, called dead clay. It is found that it is necessary to add about 20–30% new sand to the reconditioned sand in order to achieve the desired properties.

Reclamation

The reason for an amount of new sand addition is to provide clean quartz surfaces, since deterioration in bond strength occurs when the silica grains are contaminated. A clean quartz surface is of importance whether the bond is clay, sodium silicate or resin.

For clay-bonded sands the quartz grains become coated with previous clay particles by either of two following processes. The high temperature at the zones adjacent to the mould interface causes melting of the clay which flows and forms a coherent cover over the particulate matter in the form of a hard brittle shell. A shell may also form in a different way. The first return sand has already got grains coated with clay. On remilling the return sand with fresh clay, a second envelope forms around the first coating of clay. The second clay may be patchy around the first clay envelopes. Some of these second layers of clay will melt and sinter and others remain just dehydrated depending on the temperature attained by them during and after casting. On repeated recycling then, layers of clay will deposit around the prior silica grains. During further milling the brittle shells will break into fragments and, if not removed, will become embedded in the sand–clay particles during remixing with new clay giving projections on the particle surfaces. These projections decrease the flowability of the moulding mixes by increasing the apparent angularity of the particles. The coated grains increase the particle diameter. This together with the fines that are created alters the grain size of the mould or core material considerably.

In order to provide clean quartz surfaces for the development of optimum bond strength, the clay particles must be removed from the return sand, and reconditioning cannot treat the individual sand grains. Another difficulty with the use of reconditioned sand is that for each amount of new sand added to the mill for preparing the moulding material, the same amount of old sand must be removed from the foundry which adds to the sand-handling cost. These considerations have led to the methods known as sand reclamation processes devised for cleaning each individual sand grain from the knockout system free from prior bonds and additives.

There are three methods of sand reclamation, namely dry scrubbing, wet scrubbing and thermal reclamation. The last method is not used for clay-bonded sands. The principle of dry and wet scrubbing is to abrade the sand particles against each other by agitating the mix while suspended in a medium. For dry and wet scrubbing units, the respective media are air and water.

Dry Reclamation Method. The first step in any form of scrubbing unit is to break the lumpy materials and remove tramp elements by a magnetic separator. The product is then led into the scrubbing unit. The dry scrubbing principle is to agitate the sand grains in a stream of air so that the particles shot-blast each other. A complete dry scrubbing plant has been described in a previous book of this library in connection with sodium silicate bonded sands.* For clay-bonded sands the total AFS clay content in the reclaimed sand varies from 0·5% to 2·5% clay depending on the design of the plant.

Wet Reclamation. Foundries in Great Britain appear to be in agreement with the fact that the old clay shells are fragmented and these and other fines in the return sand are removed efficiently by a dry pneumatic reclamation process. However, the sand grains are known to contain depressions in their surfaces, and the clay particles anchored there cannot be removed by dry scrubbing. For clay-bonded sands it is believed that superior results are obtained by wet scrubbing the return sands.

* A. D. SARKAR, *Foundry Core and Mould Making by the Carbon Dioxide Process*, Pergamon Press, Oxford, 1964.

Wet Reclamation Method. A schematic diagram of a possible wet scrubbing unit is shown in Fig. 39. The process is controlled mainly by the scrubber which effects abrading of wet sand and the classifier where the sand grains are separated according to their sizes.

The sand is fed into the lump breaker 1 which is a tumbling barrel having perforations in one part of the body. Tumbling causes some rubbing of sand grains and, as the lumps break down, the particles pass through a screen 2 into the sand sump tank 3. Oversize

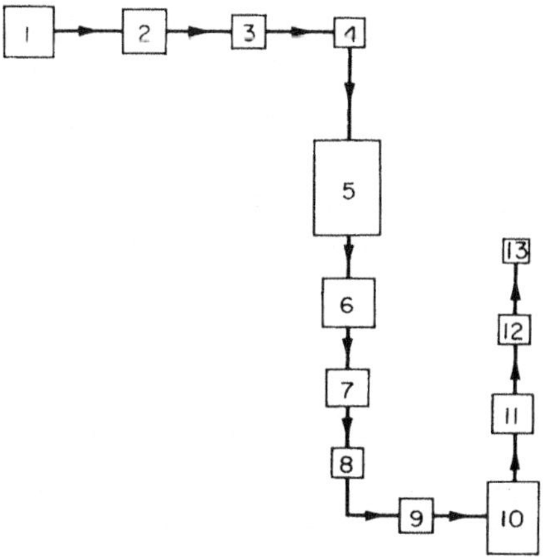

FIG. 39. Flow line of a wet reclamation system.

materials and tramp elements from the tumbler or the screen over the sump tank are pushed into a skip 4 for subsequent disposal. The slurry is pumped into the primary classifier 5 which can be a hydraulic hindered settling type and removes excessive fines from the slurry. The sand from the primary classifier is fed into the wet scrubber feed tank 6 which is primarily a storage tank, but also dewaters the slurry to 70% solids. This is the desired consistency for the most efficient wet scrubbing.

In the scrubber the slurry is thrown by an impeller which moves at 1750 rev/min. The slurry leaving at high velocity meets a slower moving stream at the circumference. The result is a very vigorous scrubbing of the sand grains against each other. The scrubbed sand is fed into a second sump tank 7 and from there pumped into the secondary classifier 8. This classifier removes all the extraneous fines and also classifies the grains. The sand is then passed through a dewatering cylinder 9. This is operated electrically and the water from the washed slurry which contains about 30% water is forced out by compressed air forcing the slurry through a screen 10. The sand leaving this dewatering filter has about 8% moisture. This is then fed into a rotary kiln type drier 11 where it is dried completely. The sand is screened 12 for the required grain size distribution and ready for the sand mill 13.

Reclaimed sands look greyish. This is because of organic matters deposited in the crevices of the sand grains. They have the effect of lowering the green strength slightly, but it is not economical to remove these by mechanical scrubbing. The organic matters can be removed by combining a thermal reclamation system with wet scrubbing, where the clay-free sands are burned in multiple hearth furnaces at a temperature of 650–750°C.

The Disposal of the Effluent Water. Wet reclamation process has found difficulty in application because of the problem of discharging the effluent water into rivers. Up to 2000 gal of effluent with solids content of about 10,000 parts per million would be produced which is about 8–10 times higher than that laid down by law. Work is in progress to find a means of clarifying the effluent so that the solids could be made to settle and the liquid recirculated in the reclamation plant. For the bentonite bonded sands addition of polyelectrolyte causes coagulation of the clay flakes. Polyelectrolyte is very expensive and will not be economical for industrial application. However, about 1% calcium chloride and polyelectrolyte in the ratio 25:1 is a very effective coagulant. The effluent water from the reconditioning plant can be treated with this when the clay settles and the supernatant liquid is inactive, that is, it is not able to coagulate the clay in the return sand further. The liquid then can be recirculated for scrubbing.

The size of flocks during the treatment with calcium chloride and polyelectrolyte depends on temperature, concentration and age of the slurry. The coagulated sludge still contains about 85% water and disposal is only possible, if it is dewatered further to reduce it to a cake which can be handled for dumping. Dewatering by centrifuging or vacuum filtration is expensive. The relatively inexpensive method appears to be rotary drum vacuum filters, the sludge being conditioned with lime before filtration.

A possible means of disposal is to thicken the slurry with waste sand. However, the sand to slurry ratio is very high, being of the order of 5:1.

CHAPTER 12

CASTING DEFECTS

DEFECTIVE castings can arise out of a variety of reasons, viz. gassy metal, improperly designed gating systems, faulty closing giving cross joints, etc. In this chapter the most common defects experienced arising out of mould and core materials are summarized. The effect of moulds and cores on scab and hot tear defects are discussed in Chapters 13 and 14.

Metal Penetration

This defect is seen on casting surfaces giving the appearance of short metal prongs which are solidified liquid steel that has penetrated into the mould and core wall through the pore spaces. It is known that metal penetration effect increases with increased ferrostatic pressure in the mould and larger pore spaces between the sand grains. There is very little that can be done to alter the ferrostatic pressure since this would depend on the depth of the mould. The only possible remedy is to diminish or seal up the pore areas by:

(1) hard ramming;
(2) using finer sand;
(3) using fillers;
(4) using a protective mould coating.

It is believed that as the steel overcomes the force of surface tension and tends to enter the pore spaces solidification of the metal also takes place and the effect of metal penetration is no more than a rough casting surface. However, if the mould interface becomes heated to a temperature where the steel would remain liquid for a long time, penetration up to the so heated depth would be expected. Use of a mould material of higher chilling power than silica should prevent the depth of penetration, and zircon is known to give a good strip with steel castings. However, the variation in chilling power between the commonly used materials in the steel foundry such as silica, zircon or chromite does not appear to be very great and it is known that coarse-grained zircon is just as prone to metal penetration as silica of comparable grain size.

Zircon and chromite as supplied are fine grained, but the materials are expensive and can only be used sparingly. The grain size of silica sand cannot be reduced indefinitely because of the possibility of increased scab defects with finer silica sand, and filler materials such as silica flour, fireclay, etc., have been used. Metal penetration is less common with machine moulded castings or those produced in moulds compacted by sand slingers. Very large jobbing work is usually rammed by hand and pneumatic rammers, and localized soft spots may be expected, being potential areas for metal penetration. A thick mould paint partitioning the metal and sand interfaces is a good safeguard.

Mould Erosion. Middleton* has made a very elaborate study of the phenomenon of mould erosion by using a test casting and has demonstrated that the larger the pore spaces in the rammed sand the more severe is the erosion. It is postulated that erosion is brought about by a physical effect where metal first gets between the pore spaces and then dislodges individual sand grains. For synthetic sands which have the sand grains distributed on a few adjacent sieves, permeability is a good parameter to assess the sizes of pore spaces. However, for the naturally bonded sands which have low permeability, it is demonstrated that because of the presence of clay and a wider sieve distribution, there will always be pockets of high permeability areas where metal penetration and dislodgement of

* *Brit. Found.* **56** (7), 337 (1963).

sand grains will occur. Erosion is a common occurrence with green-sand plate type castings. This is attributed to air drying of the surface and the material having low dry compressive strengths. It is demonstrated that the moulding material should have a dry compressive strength over 60 lb/in^2 if erosion defect is to be avoided.

An erosion effect is also experienced when the silica surfaces expand and crack ahead of the flowing metal stream, and such casting defects can be obviated by using anti-scabbing additives in the sand or using a non-siliceous material as the filler grains. The pore space can be minimized in the same ways as discussed under metal penetration. Erosion is not generally experienced with silicate–CO_2 bonded sands provided the moulds and cores are not overgassed. Silicate bonded sands have high permeability. It is often suggested that the low melting-point bond forms a ternary FeO–SiO_2–Na_2O complex which seals up the pore spaces and prevents mould erosion and metal penetration.

However, mould erosion is experienced at the points of persistent flow of metal such as at or near the ingates. This must be associated with the loss of bond at high temperature and should be a subject for further study. The ingate systems, for this reason, are lined with refractory sleeves and slabs and the practice of nailing the mould surfaces is known to prevent mould erosion.

Metal–Mould Reactions. The common defects such as blowholes and pinhole porosity are results of reaction of the steel with moisture or carbonaceous matters. In order to ensure that the steam reaction is not violent there must be a minimum amount of moisture compatible with the workability and the strength of the green-sand moulds. The usual organic binders such as oil, starch, etc., do not appear to be harmful provided the moulds and cores have reasonable permeability. Pinhole porosity and the above-mentioned defects are found to be less prevalent when the steel is efficiently de-oxidized with aluminium or other materials such as zirconium etc.

Manganese Steel. The austenitic manganese steel appears to form a reaction product at the metal–mould interface giving very poor surfaces. They often appear as irregular very rough surfaces and in many cases skins are adherent to the casting giving the appearance of thin scabs. It is difficult to grind manganese steel, and very bad surfaces may lead to scrapping the castings. Like all metal–mould reactions high pouring temperatures aggravate the surface defect. It appears that the defect is negligible with dry or silicate–CO_2 sands, but is very severe with Wyoming bentonite bonded green sand.

Colligan *et al.* have shown that the sintered layer of the mould interface shows evidence of the presence of a liquid silicate melt at the casting temperature. This melt converts essentially to fayalite upon cooling. The matrix, apart from quartz and cristobalite, shows magnetite and haematite. The mechanism is suggested as the liquid silicate melt attacking the silica and enlarging the pore spaces. It appears that the reaction occurs only at an oxidizing atmosphere and introduction of agents which might promote a less oxidizing atmosphere such as resin will minimize the defect. The resin could be applied to mould surfaces by spraying a resin solution in alcohol and then igniting it.

Shell Moulds. The two defects which are not common with other sands are orange-peel and carburization. The orange-peel defect is thought to be as a result of the expansion of the shell when the metal is in a spongy condition giving a characteristic rough surface similar to moving a trowel up vertically from damp concrete. If the expansion of the shell occurs when the metal is still molten, the orange-peel defect is not experienced. The remedy from the mould or core material point of view is to incorporate a suitable additive in the sand, and mill scale has been found very effective. The mill scale is actually incorporated in the pre-coated sand.

Carbon Pick-up. Certain Ni–Cr steels are specified with very low carbon contents for improved corrosion resistance. A typical composition is C 0·058%, Si 0·8%, Mn 0·8%, Ni 8%, and Cr 18%. Carburization takes place as a result of the volatilization of the carbon particles in the resin. The reactions are represented as follows: at the gas–steel interface:

$$2\,CO \rightleftharpoons CO_2 + C \quad \text{(in austenite).} \tag{18}$$

$$CH_4 \rightleftharpoons 2H_2 + C \quad \text{(in austenite).} \tag{19}$$

At the carbon gas interface:

$$CO_2 + C \rightarrow 2\,CO. \tag{20}$$

For carburization to occur, all the above reactions must proceed to the right, and experiments show that both reaction 19 and 20 are highly carburizing.

The carburized microsections show two distinct zones:

(1) A zone of austenite with no free carbide at the metal–mould boundary.
(2) Further from this zone there are isolated carbide particles in the austenite matrix and further in the carbides are precipitated along the grain boundaries.

It is found that the amount of carbon pick-up increases with resin content and with high casting temperatures.

Pin-hole in Air-set Cores. The urea formaldehyde base air-set binders invariably cause pin-hole porosity in steel castings and the only remedy is to use phenol formaldehyde.

Bibliography

J. M. Middleton, *Brit. Found.* **57** (1), 1 (1964).
G. A. Colligan, L. H. Van Vlack and R. A. Flinn, *Ibid.* **55** (9), 355 (1962).
J. A. Behring and R. W. Heine, *Trans. AFS*, **68**, 58 (1960).
B. K. Kaplish and H. T. Protheroe, *Brit. Found.* **54** (1), 26 (1961).

CHAPTER 13

SCAB DEFECT

SCABS and such related casting defects as rat-tails and buckles are the result of cracks appearing in the sand mould or core wall when the material is subjected to rapid heating during casting. Cracks are caused due to the rapid and considerable expansion of quartz during its transformation to the β form at the temperature of 573°C. The parts near the metal interface will expand more than those inside the mould wall because of the temperature gradient existing. If a single crack appears, the casting shows defects known as buckles or rat-tails. If the crack appears in such a manner that an area of the mould flakes off, scab defect will occur.

Experimental Studies. An exhaustive study of scab and related defects was made by Pettersson using a test casting supported by dilatometric studies of foundry sands. The moulds rammed in a jolt-squeeze machine, were 200 mm $\times$ 250 mm $\times$ 50 mm deep and were end-poured directly with steel from a 300 lb capacity high frequency furnace. The pouring rate was varied from 10 to 40 sec by controlling the sprue opening. Initially, the castings were cooled down and examined visually and the types of failure noted. In a modified experiment, the cope was raised so that its surface could be observed from the time pouring commenced. In effect, the failure of the rammed cope could not be reproduced in the casting as the drag did not touch the cope, but a quantitative evaluation of the time when the sand yielded was possible and the type of failure ascertained.

The first form of failure observed by Pettersson was falling sand grains during their expansion. The sand rain commenced from 8 to 10 sec after pouring started. This could be followed by flaking off of areas of rammed sand. Ball clay-bonded sands or materials with low moisture content rained heavily. Sodium bentonite bonded materials resisted sand rain, especially when the moulds were not soft rammed. Generally, sands possessing high hot strengths did not rain. Spalling of layers of sand usually started in 15–40 sec after casting commenced. Generally, pieces about $\frac{3}{8}$ in. thick flaked and fell off rather abruptly. In some cases, if spalling was resisted for more than 30 sec after pouring, the effect was bulging out of a layer of sand downwards which cracked and broke eventually. For sands which were still more resistant to thermal shock, cracks appeared at the edge of the mould and the heated layer failed by shear. If no yielding of the mould faces occurred after 60 sec from casting spalling did not occur subsequently.

Experiment with a Sand Slab. The work of Pettersson has been continued by Goad but by using sand slabs in the laboratory. It is suggested that the drag part of a mould will be heated the most where metal is flowing. A single metal stream would sweep the mould surface with an advancing hot front. If two or more metal streams are operative, they will meet together at a point where the mould surface is relatively cold, giving an area of thermal stress concentration which is conducive to initiation and propagation of a crack. Similar thermal environment will be experienced in the cope part of the mould as it receives radiant heat.

The principle has been simulated by heating a sand surface part of which was covered by a heat shield. The rammed sand specimen was 10 in. $\times$ 5 in. $\times$ 2 in. deep which had a central strip $\frac{3}{8}$ in. deep cut along its width. Two $\frac{1}{4}$ in. deep notches were further provided at the centre of this strip. The heat shield was a $1\frac{1}{16}$ in. wide steel strip placed $\frac{5}{16}$ in. from the face of the slab. The heat source composed of two 1 in. diameter silicon carbide rods placed $3\frac{3}{4}$ in. apart giving temperatures in the range 1300–1400°C. The parts of the sand surface receiving direct heat would expand to a greater depth than the shielded area. The effect will be for the expanded crust of sand to meet at the cool central strip from both ends giving a crack, and the notches provide the site for the crack to appear. The time when the crack appeared was noted so that a quantitative evaluation of cracking tendency was possible.

Scab Mechanism. Scab defects are more common with green synthetic sands than dry or silicate–CO_2-bonded sands. It is assumed that the moisture evaporated from the heated crust condenses behind forming a wet sand–sand interface within the mould. As the heated crust expands, detachment of the cracked piece is facilitated because of this low green strength area. It is also postulated that steam pressure builds up in this wet zone and greatly assists in the pushing out of this expanded crust. A distinction is made between riserless moulds and those with risers. A riser assists steam to escape during the early part of pouring when the metal is heating the cope by radiant heat. In the case of blind moulds, the steam generated condenses in a position behind the crust which is being dehydrated.

The present state of knowledge regarding the mechanism of scab formation is shown schematically in Fig. 40. Radiant heat from the rising metal causes sand rain as a first effect of failure in the rammed cope (Fig. 40a). Steam escapes behind the heated crust and condenses in a plane forming a line of weakness (Fig. 40b). Expansion causes a crack (Fig. 40c) which then bulges downwards allowing metal to enter forming a scab (Fig. 40d). If the crack is such that metal does not penetrate behind, the defect becomes a buckle or rat-tail. When metal dislodges the sand, a scab forms. The cracking and flaking off of layers of sand may also occur before metal touches the cope surface, again giving rise to scab defects.

Possible Remedy for Scab Defect. In view of the expansion of silica, it is necessary to have an optimum amount of clay in the mixes when using clay-bonded sands. An adequate amount of clay minimizes contact between the expanding sand grains and provides a cushioning effect. It is also suggested that the thin wet zone behind the heated crust will not lead to falling off of layers of sand if a clay is used which has a degree of green strength even in a highly wet condition. The Wyoming bentonite is the least water sensitive among the various clays used in connection

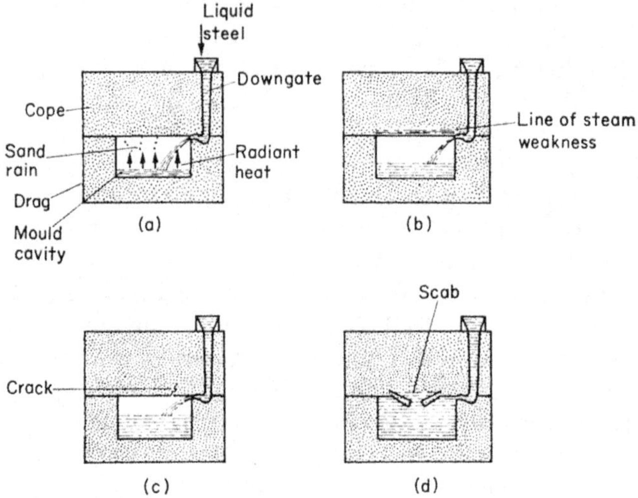

FIG. 40. Stages in scab defect (schematic).

with steel foundry mould or core materials. Wyoming bentonite is known to offer the most resistance to scabbing defect.

Work done by Parkes and Godding has shown that hot compacted sands should show some plasticity. For silica sands, cracks will occur giving scab defects if the expansion due to $\alpha \rightleftharpoons \beta$ quartz transformation exceeds deformation. Hot plasticity can be increased by an addition of:

(1) Cellulose materials.
(2) Cereal binders.
(3) Bituminous coal and pitch.

It is also important to work in the low moisture range, to use high clay content and to limit the amount of fines in the mixes. This is specially important with high pressure moulding where a squeeze pressure as high as 200 lb/in^2 may be applied. Scabbing can also be avoided by using non-siliceous materials. A high pouring temperature is desirable.

An account of research to date on scabbing and rat-tailing behaviour of foundry sands has been given recently. Foundry laboratories could usefully employ a simple equipment to measure the "wet tensile strength" of green sands where the moisture condensation is allowed to occur several millimetres away by heating the face of a standard specimen and breaking it in tension. A typical sand which may give a minimum of rat-tailing in a casting has a wet tensile strength of the order of 10 g/cm^2.

Whether hot strength characteristics of foundry sands can predict with certainty the scabbing propensity of a rammed mould has to be evaluated. Possibly, the parameters to measure are the peak hot strength temperature and the maximum load-bearing capacity. An interesting point is that, during the application of compressive loads on hot samples, in some cases specimens fail by shear while in others failure is preceded by subsidence, that is, there is a degree of deformation before failure. Subsidence means that the bond is melting. It is worth noting that the silicate–CO_2- and sodium montmorillonite-bonded sands show less scabbing tendencies than the high melting-point clays. However, it must also be noted that the silicate–CO_2-bonded sands are very nearly moisture free, while the sodium bentonites are less water sensitive.

Bibliography

H. Pettersson, *Foundry Trade Jl.* **99**, 263, 301 (1955).
P. W. Goad, *Trans. AFS,* **67**, 436 (1959).
J. A. Kilshaw, *Jl. BCIRA*, **12**, 14 (January 1964).
F. Hofmann, *Trans. AFS*, **70**, 1 (1963).
F. Hofmann, *Brit. Found.* **52**, 196 (1959).
W. B. Parkes and R. G. Godding, *Foundry Trade Jl.* **99**, 139 (1955).
H. W. Dietert and T. E. Barlow, *Trans. AFS*, **66**, 7 (1958).
A. D. Morgan, *Brit. Found.* **59**, 186 (1966).

CHAPTER 14

HOT TEAR

BRIGGS and Gazelius have identified two types of cracks which may occur in a steel casting while it solidifies in a mould, viz. hot cracks and cold cracks. Both types arise from contraction stresses in the casting, these stresses developing as a result of volume change following the transition from the liquid to the solid state and during cooling from the solidus to ambient temperature. In a freely contracting bar there are no stresses due to contraction, and those during cooling are only developed if the volume change is resisted. Resistance to volume change is the result of casting design and/or surrounding mould–core material. The main effect of casting design is non-uniform cooling whereby a thin solid section tends to pull away from adjacent relatively liquid thick section, and can be obviated by suitably positioning ingates, risers and using chills or mould materials of differing heat extracting power to ensure directional solidification.

Hindrance to contraction may also be offered by a mould or core material. For instance, contraction of a ring casting may be hindered by the core and stresses arising due to this hindrance may be relieved by the formation of a crack at some part of the casting. If this stress relief takes place at relatively low temperatures the cracking is designated cold cracking. When cracks occur in a steel casting at high temperatures (of the order of 1300°C) they have the appearance of a tear and because of the jagged appearance of the fracture this type of cracking has been designated as hot tear. There are two types of hot tear, viz.:

(a) Internal hot tear.
(b) External hot tear.

Internal hot tears unlike external hot tears seldom appear on the casting surface and are a result of casting design which causes non-uniform cooling. They appear characteristically as groups of small cracks and may also be initiated by small shrinkage cavities which act as areas of stress concentration.

External hot tears are attributed wholly to the resistance offered by moulding materials and are so termed because they originate on the outside of the component and are easily detected on rough castings. In this chapter only the phenomenon of external hot tear and the role of mould material on this defect will be considered although occurrence of hot tear may also depend on the steel composition, casting temperature and deoxidation practice.

Resistance by a Mould–Core. Resistance to contraction of a casting may be offered by the rammed mould or core material, and a number of investigators have made attempts to correlate properties of foundry sands with the susceptibility to hot tears of steel castings.

The steel research committee of American Foundrymen's Society have used a variety of castings to study the effect of mould materials on the susceptibility of steel castings to hot tear. First a 4 in. high bushing with $\frac{1}{2}$ in. metal thickness was used, the inside diameter being varied from 1 in. to 6 in. It was concluded that hot tearing increased with increasing diameter of the bushing. Although there was a considerable scatter between results, the general tendency was for increased bulk density of the core to offer greater hindrance to the contracting diameter, and hot tearing was accentuated. The bushing in this study was cast vertically. It was later decided to use a similar casting but cast with its axis horizontal. Metal was introduced at the underside of the bushing by means of a slot gate which ran almost the whole length of the casting. The conclusions were that:

(a) Hot tear will occur at hot spots in a casting.
(b) Tendency to tearing increases with increased bulk density of rammed moulds and cores.

Middleton has made an attempt to evaluate quantitatively the incidence of hot tearing by using a test casting as shown in Fig. 41. The effect of various types and amounts of bond, ramming density and grain size was studied using two castings weighing 10 and 35 lb respectively. The small casting had a cross-sectional area of 1 in. × 1 in. and was 21½ in. long, while the large casting was 2 in. × 1 in. × 38½ in. The bar was moulded by using a two part box and the notched bolt was placed in the mould cavity in order that when metal was introduced into the mould the bolt would grip the solidifying casting. The other end of the bolt was

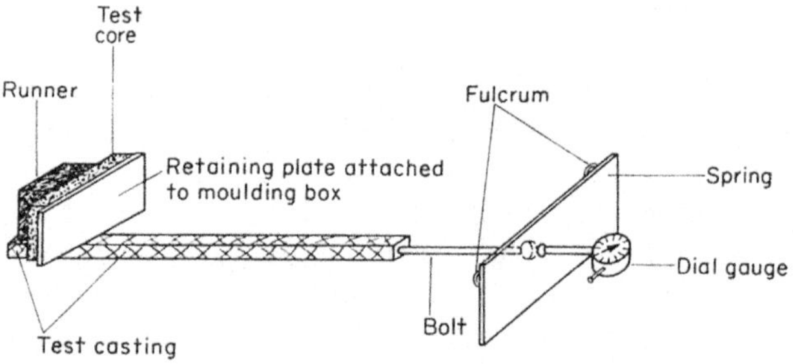

FIG. 41. Hot tear test casting (Middleton).

attached to a plunger which jointed onto a flat spring. The spring was designed to deflect 0·001 in. for each load increment of 10 lb. The bending moment of the spring was read from a calibrated dial gauge. The casting was well fed. The junction of the runner and the casting created the hot spot where tearing would occur. An amount of hindrance to contraction was offered by the bolt and spring and further resistance to contraction was allowed for by placing the rammed material under study at the junction of the hot spot. The test core was placed so that one face of it formed the inside vertical wall of the runner and the other face rested against a steel plate which was secured to the top part moulding box. This meant that as the runner tended to contract towards the direction of the bolt, its movement would be restricted unless the rammed core deformed and collapsed. That is, if the rammed core had high compressive strengths under those temperature conditions tearing at the hot spot would be expected.

The blocks were $5\frac{1}{2}$ in. × 4 in. × $2\frac{1}{2}$ in. thick and were made by compressing a known weight of sand in a core box by using a tensile machine. The bulk density of the sands could be varied by altering the weight of sand in the core box. Melting conditions were kept constant between tests. The load, as measured by the spring, was taken as the criterion of severity of the moulding material regarding hot tear, a higher load meaning higher resistance to the contracting casting. The conclusions from a mild steel were that increased resistance to contraction and hence increased likelihood of hot tearing is experienced when

(a) the core is made in dry sand; and when
(b) the bulk density of sands increases.

The tendency to tearing is less with bentonite-bonded sands and more with kaolinitic clay, the most severe tearing being experienced with a mix containing fireclay, bentonite and 20% silica flour.

The effect of increasing the bulk of the casting was noticeable in the case of organic binders. For, whereas the small castings tore severely with oil-bonded sands, the tear was minimized considerably when the larger casting was used. Slightly larger tears were associated with phenol and urea resins. In a separate survey Middleton has reported that the propensity to hot tear is no greater with sodium silicate–CO_2-bonded sands than with clay-bonded sands. However, there is no general agreement on this conclusion among foundrymen.

The role of certain elevated temperature properties in connection with external hot tear can, possibly, be interpreted in the following way. If a casting with a cube core is considered, upon the introduction of metal, the core interfaces will be heated to some subsolidus temperature, and points further removed from the interface will have progressively lower temperatures. For the casting to contract inwards the cube core must compress in all directions. A property of bonded sands is that they are dilatant, i.e. under compressive force the grains tend to slide past one another and repack to a higher density. There is a maximum density per unit volume to which a particular mould material can be compressed. The core, since it is rammed, will already have a degree of inherent compressive strength. That is, it will resist an amount of external load before it deforms and ultimately ruptures. For deformation to occur the stress induced by the contracting casting must be greater than the compressive strength of the rammed core material thus allowing the core to deform or collapse and removing the resistance to contraction. It is known that the compressive strengths of bonded sands vary according to the temperature at which the rammed specimens are soaked. Since various layers of the core will be at different temperatures it will be the effect of the hot strengths or deformability of one or more layers that will decide whether a core or mould will offer the amount of resistance for tear to occur. There is a further parameter to be considered. This is the retained strength of the sands at a high temperature after being soaked to some higher temperatures previously. It is known that the parts of the mould–core near the interface tend to rise quickly and to fall to some lower temperature in a short period of time. It is possible that contractional stresses in a casting develop only at that later stage and it would be the retained strength of the sand at that temperature that should determine the behaviour of that layer of sand.

Some explanations are offered from the work of Middleton regarding the hot properties of foundry sands. For example, bentonite led to less tear than the fireclay and it is known that the former is a lower melting-point bond than the latter. Whenever a bond melts, a marked plasticity results and the sand deforms under compressive loads. With oil-bonded sands, the result is a physical effect where the oil burns leaving pore spaces for the granular particles to move allowing the core to collapse. The small casting tore severely while the large casting was free from cracks. The most probable reason is that because of the mass effect the heat was sufficient to cause burning of oil in the large casting and, in the small casting, the heat was inadequate.

References

C. W. Briggs and R. A. Gazelius, *Trans. AFA*, **41**, 385 (1933).
C. W. Briggs, *Trans. AFA*, **45**, 61 (1937).
G. W. Johnson, *Trans. AFA*, **60**, 137 (1952).
G. A. Killiequist and J. H. Fuqua, *Ibid.* p. 141.
C. H. Wyman, C. A. Faist and G. Di Sylvestro, *Ibid.* p. 145.
C. Wyman, *Trans. AFS*, **58**, 144 (1950).
J. M. Middleton, *Trans. AFS*, **61**, 167 (1953).
J. M. Middleton, *Trans. AFS*, **63**, 99 (1955).
J. M. Middleton, *Brit. Found.* **53**, 67 (1960).

CHAPTER 15

HIGH TEMPERATURE PROPERTIES

THE first action of liquid steel during the casting process is to subject the mould and core walls to intense thermal shock. The next stage is for the mould interface in contact with the metal to be heated to the near solidus temperature of the steel while planes successively behind will be at progressively lower temperatures, i.e. the mould or core has a temperature gradient across the width of its wall.

The heated mould is also under compressive stresses due to the static load imparted by the box parts and the ferrostatic pressure of the steel. The vertical planes of the mould should tend to bulge out and fail if the compressive and the tensile strengths of the material are inadequate at the temperatures prevailing. That is, a knowledge of bond strength of a mould material at high temperatures is necessary. For example, during the dynamic flow of metal in the mould if the cohesive force between the particulate matters is diminished mould erosion would be expected particularly at the points of persistent flow such as at or near the ingates. The bond strength may also be inadequate when a solid skin has already formed on the casting when the erosion effect will not be expected. However, the particulate matters may detach from the mould faces and deposit on the solid, but hot casting surface giving, possibly, the phenomenon of burnt-on sand.

Hot Strength

Hot compressive strengths have been measured by various workers. In isothermal heating a specimen is heated until the temperature gradient from its outside to the centre evens out, and then broken under compression. It is known that the hot compressive strength increases with increasing temperature to a maximum and falls thereafter. The peak temperature depends on the type of clay. The sodium montmorillonite gives the highest peak hot strength at the lowest temperature, while kaolinite shows a low strength magnitude occurring at a temperature of the order

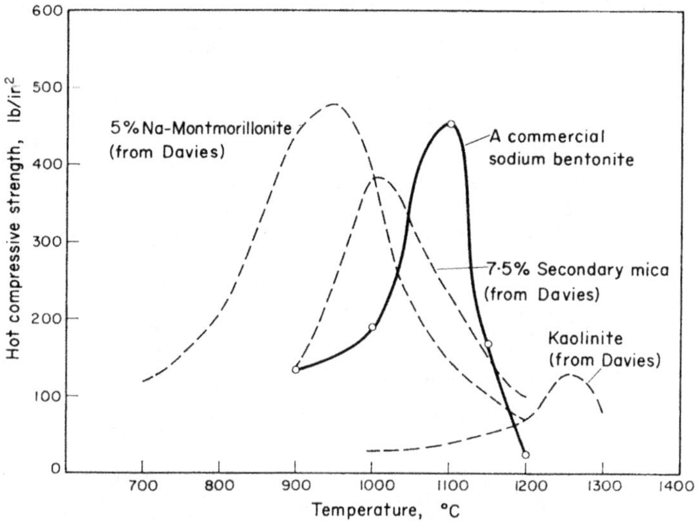

FIG. 42. Isothermal strength curves.

of 1250°C (Fig. 42). The secondary mica occupies an intermediate position, while a commercial bentonite used by the present author shows a peak strength comparable to the sodium montmorillonite but occurring at some 150°C higher and possibly indicates the presence of impurities. Most of the work on hot compressive strengths have been carried out along similar lines to Davies and Rees and such variables as amount of clay, ramming densities, effect of additives, etc., have been studied.

The present author has continued the study of high temperature

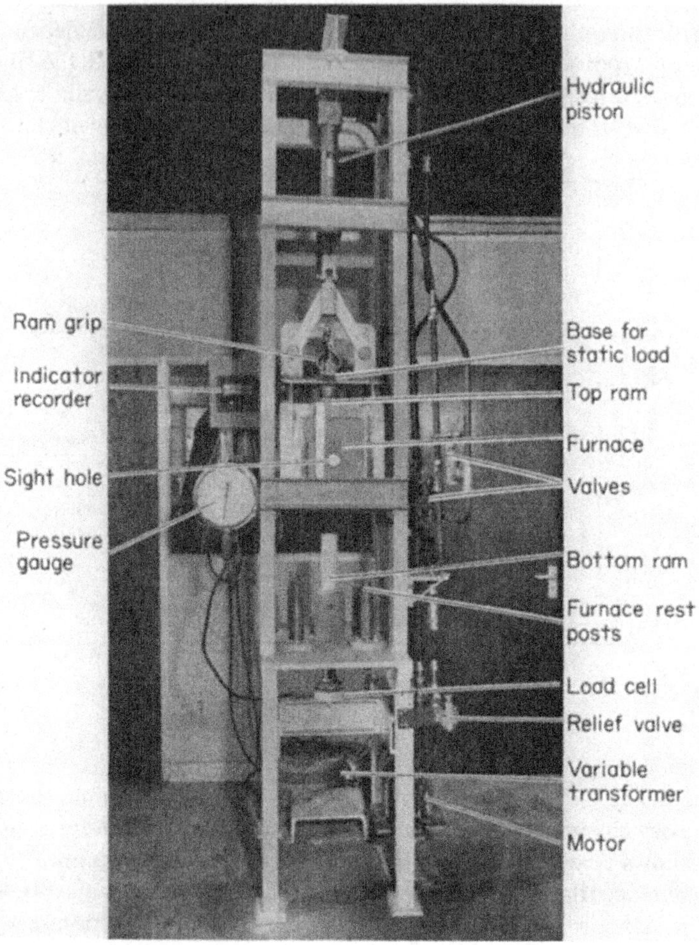

FIG. 43. Hot strength apparatus.

properties by using a furnace (Fig. 43) capable of moving in a vertical direction, the object being to place the rammed-sand specimen on the bottom ram and quickly bring the furnace, held at a predetermined temperature, down on the sample so that the shock heating condition such as it prevails in a mould can be simulated. The resistance to thermal shock was studied qualitatively by using a domed-top AFS compressive specimen and noting the time when the first and subsequent cracks appeared in the specimen. The hot compressive strengths were plotted against various soaking times at constant temperatures. The conclusions were generally similar to those of other workers, viz. hot compressive strengths increased with increased ramming densities or when certain fillers such as zircon flour were incorporated in the mould or core materials. It appeared significant, however, that the strength–time curves could be divided into two stages. In the early stages, the condition is layer heating where a definite temperature gradient exists between the outside and the centre of the specimen. This is precisely the condition that exists in a mould and core when the metal introduced begins to heat up the interfaces so that the stresses are acting on a volume which can be regarded to be generated by various layers of material possessing differing inherent strengths depending on the thermal history of them. When the soaking time of the hot strength specimen is long enough to even out the temperature gradient between the outside and the centre, the sand sample could be regarded as the plane of a mould or core at that temperature. For example, if the furnace temperature employed was that of the metal–mould interface, the hot bond strength of the mould interface could be evaluated from a knowledge of the hot compressive strength of a specimen broken after heating isothermally.

Data from some production sands on high temperature properties are shown in Chapter 17, and the possible role of hot strengths or resistance to thermal shock effect of steel foundry mould materials on cast quality is indicated. However, precise conclusions regarding the role of high temperature properties, especially the hot compressive strengths, on cast quality are still lacking.

Bibliography

W. Davies and W. J. Rees, *Jl. ISI*, **152**, 61P (1946).
R. W. Heine, J. S. Schumacher, W. F. Shaw and E. H. King, *Trans. AFS*, **68**, 304 (1960).
R. W. Heine, J. S. Schumacher and E. H. King, *Trans. AFS*, **69**, 504 (1961).
N. C. Howells, R. E. Morey and H. F. Bishop, *Trans. AFS*, **65**, 402 (1958).
A. D. Sarkar, *Foundry Trade Jl.* **119**, 671 (1965).

CHAPTER 16

TECHNICAL CONTROL

IN ORDER to produce mould and core materials giving consistent green, dry and hot properties, it is imperative that the raw materials conform to a standard specified by the foundry. The raw materials are broadly sand, binder and additives.

Sand. Silica sand is used in bulk in most steel foundries either for preparing synthetic mixes or as diluents to natural sands containing excessive clay and poor refractoriness. For steel foundry application, the silica sand should contain at least 96% SiO_2 and less than 1% alkalis such as Na_2O and K_2O. The alkalis are present in the form of feldspar which has a low fusion point, being of the order of 1250°C resulting in a low refractory moulding material. A complete chemical analysis of the sand is unnecessary and generally the silica and the alkalis only need to be determined.

The sand must then be checked for sieve distribution. Typical particle size distribution of a coarse, medium and fine sand used successfully in steel foundry application is shown in Table 21. The AFS clay content and the sieve analysis of a naturally bonded sand is also included. Bad casting surface by metal penetration or mould erosion may be experienced with too coarse a sand, while fine sands tend to produce scab defects. Excessive fines such as the particles passing through 300 mesh will not only decrease the permeability of the mixes, but would necessitate an increased amount of bond for optimum property development. Excess bond is uneconomical and, in the case of oil bond, the amount of gas evolved from the core material will be high as the quantity of binder is increased thus increasing the probability of gas holes in the casting since the fines will also decrease permeability.

TABLE 21. SIEVE DISTRIBUTION OF SANDS (% RETAINED)

BS sieve No.	Silica sand			Natural sand[1]
	Coarse	Medium	Fine	
8	–	–	nil	nil
10	–	–	nil	0·1
16	0·1	0·3	nil	0·8
22	0·4	1·0	nil	2·2
30	2·1	4·2	nil	4·7
44	16·4	15·1	0·5	16·5
60	40·2	33·4	6·0	26·5
72	27·3	19·7	10·5	12·9
100	12·9	17·2	46·0	10·4
150	0·55	8·2	27·5	4·5
Pan	0·55	0·9	9·5	5·9

[1] % AFS clay = 15·5.

A difficulty with dry silica sand is that the grains tend to segregate easily during transport and storage. This means that grain-size distribution is altered when sand is withdrawn from the storage hoppers. Apart from the limitations imposed as described in the previous paragraph, a badly graded sand is known to be thermally unstable which would cause the rammed mould and core to crack and spall more easily than when a properly graded sand is employed. The segregation difficulty could be avoided by slightly dampening the sand during transport and storage. However, certain non-clay bonded sands such as thermosetting and air-set resin bonded mixes require silica sands without even a trace amount of moisture. Some recommendations are that during charging the sand hoppers the sand should not be allowed to fall through a great height. The sand should be withdrawn from a large number of points in the storage bunkers for charging the sand mill weigh hoppers.

The first variable to be assessed with naturally bonded sands is its clay content since this varies between sources of the sand and even from batch to batch mined from the same quarry. In certain deposits the clay exists as an interlayer between thicker layers of quartz. If this clay, after mining, is not broken up and distributed evenly among the quartz particles, clay lumps will remain segregated in the natural sand when received in the foundry. These clays apart from being less refractory than the silica grains hold excessive moisture and may cause localized blowing during and after casting. Naturally bonded sands must therefore be milled thoroughly in a roller mill to ensure breaking up of all clay lumps and their subsequent distribution. Apart from the clay content, a refractoriness at 1550°C should be carried out. If the per cent linear change is negative, an amount of silica sand should be added. Often two or more naturally bonded sands are blended together giving a suitable moulding material. The AFS clay content is held at about 13% and the moisture is of the order of 7%.

Clays. The properties such as gelling index, base exchange capacity, etc, as described in Chapter 5, should be measured. However, the most rapid and useful test for a clay is to compare its green and dry properties together with the refractoriness at 1550°C with a standard clay. For example, with montmorillonites a Wyoming bentonite is used as a standard. A quantity of the standard clay and clean silica sand of medium grain size is stored in drums in the laboratory. A table such as Table 22 is prepared where a comprehensive range of property variation with varying amounts of water and bentonite are recorded. Usually, three ram specimens are used, but for certain clays of low green strength six or nine ram specimens are also used. For rapid evaluation of the suitability of the clay received, a 5% clay and 4% water is mixed in the laboratory mill with silica sand and the properties obtained compared with the standard as in Table 22, which can also be represented graphically. If the clay seems promising a more comprehensive test programme as in Table 22 should be undertaken and the amount of clay and water optimized.

TABLE 22. GREEN AND DRY PROPERTIES

% Wyoming bentonite	% water	Green compressive strength (lb/in²) at		
		3 rams	6 rams	9 rams
4	4	4·5	5·3	6·1
4	5	4·6	–	–
4	6	4·0	–	–
6	4	7·0	8·4	10·5
6	5	7·5	–	–
6	6	6·5	–	–
8	4	9·8	13·6	15·3
8	5	9·9	–	–
8	6	8·5	–	–

% Wyoming bentonite	% water	Dry compressive strength (lb/in²) at		
		3 rams	6 rams	9 rams
4	4	91·00	121·00	116
4	5	106·50	–	–
4	6	108·50	–	–
6	4	128·00	157·30	154
6	5	139·50	–	–
6	6	134·00	–	–
8	4	150·00	174·50	194
8	5	157·00	–	–
8	6	158·50	–	–

% Wyoming bentonite	% water	Green perm. No. at			Shatter index % at 3 rams
		3 rams	6 rams	9 rams	
4	4	165	118	110	69
4	5	142	–	–	72
4	6	137	–	–	69
6	4	168	110	88	89
6	5	141	–	–	91
6	6	110	–	–	90
8	4	157	105	80	99
8	5	142	–	–	99
8	6	125	–	–	99

Durability Test. Since in the return foundry sand an amount of the prior moulding material will lose the clay bond which becomes dead clay by virtue of its loss of constitutional water, a check on the life of a clay is a useful guide in assessing the amount of new

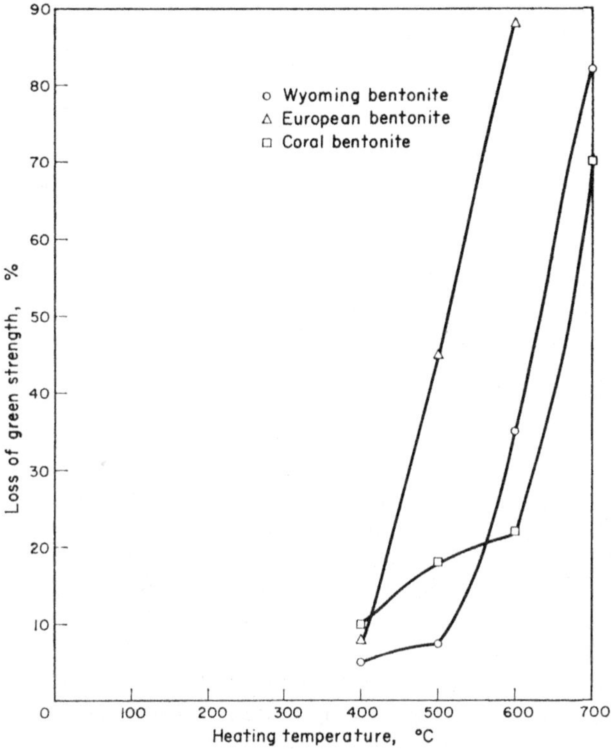

(*Courtesy Sternol Ltd.*)

FIG. 44. Durability of clays.

bond to be added to the sand system. This is done by milling silica sand, for example, with 5% clay and 3% water and noting the green strength. The sand–clay mixes are heated to various temperatures, cooled and remilled to give a final moisture of 3% in the material. The green strengths from each mix are measured and the per cent loss of green strength in terms of the original sand is expressed against various heating temperatures. Graphs are plotted as in Fig. 44. Here, for example, the coral bentonite has lost only 23% of its original strength at 600°C and is superior to the Wyoming bentonite and European bentonite from the durability point of view.

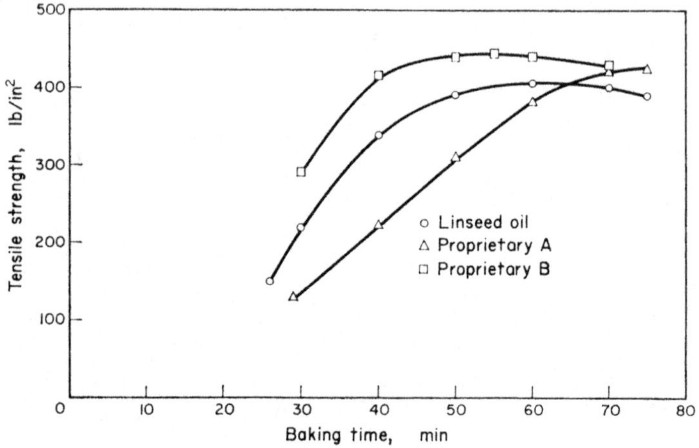

(*Courtesy Harborough Construction Company Ltd.*)

FIG. 45. Baked strengths of oil sands.

Oil Bonds. The various proprietary oil-based binders and core creams conform to a definite standard between suppliers. If time and staff are available, the saponification number should be evaluated. However, a more direct test is to have a comparison with linseed oil by using the baked tensile strength at a constant baking temperature and time as the parameter for comparison. Figure 45 shows the variation in drying rate of two proprietary core oils. Both give optimum baked strength with the difference that one develops strength at a faster rate than the other, that is the former is a fast-drying oil.

Of the other types of bonds and additives, usually suppliers' data regarding their properties are accepted. However, comparative tests can always be made with a standard mix which has proved to be successful in the foundry. New materials not used hitherto in the foundry should be investigated in the non-routine section of a laboratory, and conclusions regarding the suitability of the product should be drawn on the basis of the results obtained with experimental and production castings.

Control Charts

Provided the basis materials are under proper control the product from the sand mill is generally of consistent properties for oil-, silicate- and air-set bonds. However, the green facing or backing sands and especially the unit sands are very difficult to control. The main reason for this is the uncertainty in the return sand. This may arise out of contamination with other bonds such as silicate, oil, etc., and further the amount of dead clay or fines due to cracking of the sand grains during alternate heating and cooling arising out of the casting processes will vary. A sand reclamation plant incorporated in the return sand system minimizes these variables largely, but even then an amount of new sand is added with reclaimed sand to provide bonding surfaces compensating the parts of the sand grains which have not been cleaned because of the physical nature of the grains. Other variables such as mill wear or error in the weigh hoppers, etc., are inherent sources of property fluctuations in moulding mixes and therefore routine property measurements on each mix must be carried out. It is desirable to have a check on all the properties and the results are best expressed in the form of control charts which allow the sand technician to detect any dangerous deviation from a norm on a day-to-day notice. A control chart of green compressive strength and that of moisture variation is shown in Figs. 46 and 47 respectively.

Plotting of control charts necessitates the following observations over a period of time (Tables 23 and 24).

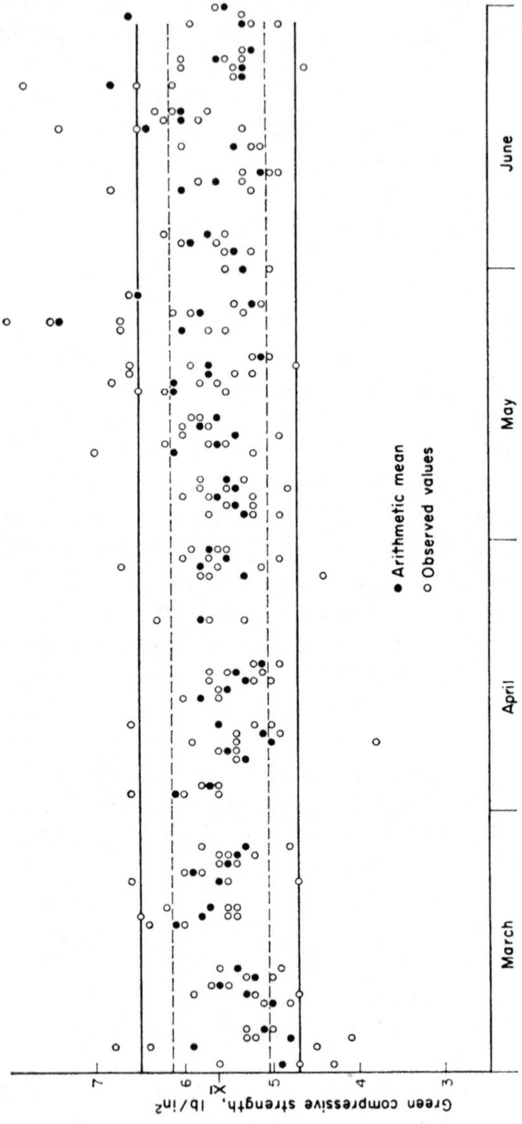

FIG. 46. Control chart for green compressive strength.

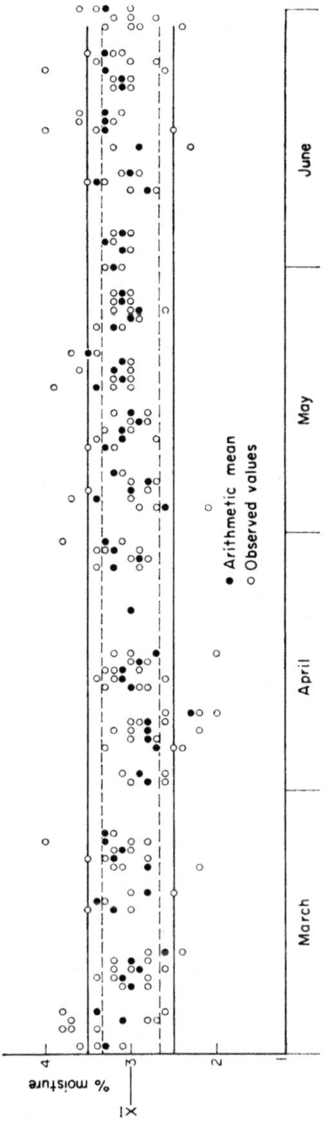

FIG. 47. Control chart for moisture.

TABLE 23. GREEN COMPRESSIVE STRENGTH (lb/in^2)

March	Samples	Mean/Range[1]	March	Samples	Mean/Range
1st	5·6 4·7 4·3	(1·3) *4·9*	17th	6·4 6·0 6·0	(0·4) *6·1*
3rd	4·5 6·4 6·8	(2·3) *5·9*	18th	6·5 5·5 5·4	(1·1) *5·8*
4th	5·3 4·1 5·2	(1·2) *4·8*	19th	6·2 5·5 5·4	(0·8) *5·7*
5th	5·3 5·1 5·0	(0·3) *5·1*	22nd	6·6 4·7 5·5	(1·9) *5·6*
8th	5·1 4·8 5·1	(0·3) *5·0*	23rd	6·0 5·8 6·0	(0·2) *5·9*
9th	4·7 5·2 5·9	(1·2) *5·3*	24th	5·6 5·6 5·4	(0·2) *5·5*
10th	5·7 5·5 5·6	(0·2) *5·6*	25th	5·6 5·5 5·2	(0·4) *5·4*
11th	5·2 5·3 5·0	(0·3) *5·2*	26th	4·8 5·3 5·8	(1·0) *5·3*
12th	5·6 5·6 4·9	(0·7) *5·4*			

[1] Mean values in italic. Range values, *w* in ( ).

TABLE 23. (*contd.*)

April	Samples	Mean/Range	April	Samples	Mean/Range
1st	6·6 5·6 6·0	(1·0) *6·1*	14th	5·2 5·0 5·7	(0·7) *5·3*
2nd	5·6 5·7 5·8	(0·2) *5·7*	15th	5·1 5·7 5·5	(0·6) *5·4*
5th	5·4 5·3 5·3	(0·1) *5·3*	16th	4·9 5·2 5·1	(0·3) *5·1*
6th	5·4 5·6 5·4	(0·2) *5·5*	21st	6·3 5·7 5·3	(1·0) *5·8*
7th	5·9 3·8 5·4	(2·1) *5·0*	26th	4·4 5·7 5·8	(1·4) *5·3*
8th	4·9 5·4 5·1	(0·5) *5·1*	27th	5·1 5·6 6·7	(1·6) *5·8*
9th	5·2 6·6 5·0	(1·6) *5·6*	28th	4·9 6·0 5·7	(1·1) *5·5*
12th	5·8 5·6 6·0	(0·4) *5·8*	29th	5·6 5·9 5·5	(0·4) *5·7*
13th	5·5 5·5 5·6	(0·1) *5·5*			

TABLE 23 (*contd.*)

May	Samples	Mean/Range	May	Samples	Mean/Range
3rd	5·7	(0·8)	18th	5·8	(1·2)
	5·2			6·8	
	4·9	*5·3*		5·6	*6·1*
4th	5·5	(0·3)	19th	5·4	(1·4)
	5·4			5·2	
	5·2	*5·4*		6·6	*5·7*
5th	6·0	(0·8)	20th	5·9	(1·9)
	5·2			6·6	
	5·7	*5·6*		4·7	*5·7*
6th	4·8	(1·0)	21st	5·0	(0·2)
	5·8			5·1	
	5·5	*5·4*		5·2	*5·1*
7th	5·3	(0·5)	24th	5·5	(1·2)
	5·5			6·7	
	5·8	*5·5*		5·7	*6·0*
10th	5·2	(1·8)	25th	6·7	(1·3)
	6·1			8·0	
	7·0	*6·1*		7·5	*7·4*
11th	5·0	(1·2)	26th	5·9	(0·8)
	6·2			5·3	
	5·7	*5·6*		6·1	*5·8*
12th	4·9	(1·1)	27th	5·2	(0·3)
	5·4			5·1	
	6·0	*5·4*		5·4	*5·2*
13th	5·8	(0·3)	28th	6·6	(0·1)
	6·0			6·5	
	5·7	*5·8*		6·5	*6·5*
14th	5·2	(0·6)	31st	5·5	(0·5)
	5·9			5·5	
	5·8	*5·6*		5·0	*5·3*
17th	6·2	(1·0)			
	5·5				
	6·5	*6·1*			

TABLE 23 (*contd.*)

June	Samples	Mean/Range	June	Samples	Mean/Range
2nd	5·5	(0·3)	21st	6·1	(1·7)
	5·2			6·5	
	5·4	*5·4*		7·8	*6·8*
3rd	5·6	(0·4)	22nd	5·4	(0·1)
	6·0			5·3	
	6·0	*5·9*		5·3	*5·3*
4th	6·2	(0·7)	23rd	5·4	(1·4)
	5·5			6·0	
	5·5	*5·7*		4·6	*5·3*
9th	5·2	(1·6)	24th	5·5	(0·7)
	6·0			6·0	
	6·8	*6·0*		5·3	*5·6*
10th	5·8	(0·5)	25th	5·2	(0·1)
	5·6			5·2	
	5·3	*5·6*		5·3	*5·2*
11th	5·3	(0·4)	28th	4·9	(1·0)
	4·9			5·2	
	5·0	*5·1*		5·9	*5·3*
14th	6·0	(0·9)	29th	6·6	(2·5)
	5·2			7·8	
	5·1	*5·4*		5·3	*6·6*
16th	5·3	(2·1)	30th	5·6	(0·1)
	7·4			5·5	
	6·5	*6·4*		5·5	*5·5*
17th	6·2	(0·4)			
	5·8				
	6·0	*6·0*	$\bar{x} = 5{\cdot}6$		
18th	6·1	(0·6)			
	6·3				
	5·7	*6·0*	$\bar{w} = 0{\cdot}85$		

TABLE 24. % MOISTURE IN GREEN SAND

March	Samples	Mean/Range[1]	March	Samples	Mean/Range
1st	3·1 3·6 3·4	(0·5) *3·3*	17th	3·0 3·5 3·2	(0·5) *3·2*
3rd	3·4 3·8 3·8	(0·4) *3·7*	18th	3·4 3·3 3·4	(0·1) *3·4*
4th	2·8 3·7 2·7	(1·0) *3·1*	19th	2·5 2·8 3·0	(0·5) *2·8*
5th	2·6 3·8 3·8	(1·2) *3·4*	22nd	2·2 3·2 3·1	(1·0) *2·8*
8th	3·1 3·1 2·8	(0·3) *3·0*	23rd	2·8 3·3 3·5	(0·7) *3·2*
9th	2·8 3·4 3·2	(0·6) *3·1*	24th	3·0 3·2 3·2	(0·2) *3·1*
10th	3·2 3·0 2·6	(0·6) *2·9*	25th	2·8 3·0 4·0	(1·2) *3·3*
11th	2·8 3·0 3·2	(0·4) *3·0*	26th	3·3 3·2 3·3	(0·1) *3·3*
12th	2·8 2·4 2·6	(0·4) *2·6*			

[1] Mean values in italic. Range values, *w* in ( ).

TABLE 24 (*contd.*)

April	Samples	Mean/Range	April	Samples	Mean/Range
1st	2·6	(0·4)	14th	3·3	(0·4)
	3·0			2·9	
	2·8	*2·8*		3·2	*3·1*
2nd	2·6	(0·5)	15th	3·0	(0·2)
	3·1			3·0	
	2·9	*2·9*		2·8	*2·9*
5th	3·3	(0·9)	16th	2·0	(1·2)
	2·5			3·2	
	2·4	*2·7*		3·0	*2·7*
6th	2·8	(0·3)	21st	3·0	(0·0)
	3·0			3·0	
	2·7	*2·8*		3·0	*3·0*
7th	2·2	(1·0)	26th	2·9	(0·5)
	3·0			3·4	
	3·2	*2·8*		3·2	*3·2*
8th	3·0	(0·4)	27th	3·0	(0·2)
	2·9			2·8	
	2·6	*2·8*		2·9	*2·9*
9th	2·6	(0·6)	28th	2·9	(0·5)
	2·2			3·4	
	2·0	*2·3*		3·3	*3·2*
12th	3·3	(0·5)	29th	3·1	(0·7)
	2·9			3·1	
	2·8	*3·0*		3·8	*3·3*
13th	3·4	(0·8)			
	2·6				
	3·2	*3·1*			

TABLE 24 (*contd.*)

May	Samples	Mean/Range	May	Samples	Mean/Range
3rd	2·1	(0·8)	18th	3·0	(0·2)
	2·7			3·2	
	2·9	*2·6*		3·1	*3·1*
4th	3·0	(0·7)	19th	3·1	(0·6)
	3·4			3·6	
	3·7	*3·4*		3·0	*3·2*
5th	3·5	(0·7)	20th	3·0	(0·1)
	2·8			3·1	
	2·8	*3·0*		3·1	*3·1*
6th	2·7	(0·3)	21st	3·4	(0·3)
	3·0			3·5	
	2·7	*2·8*		3·7	*3·5*
7th	3·2	(0·1)	24th	3·4	(0·3)
	3·1			3·2	
	3·2	*3·2*		3·1	*3·2*
10th	3·2	(0·3)	25th	3·0	(0·1)
	3·2			2·9	
	3·5	*3·3*		3·0	*3·0*
11th	2·7	(0·7)	26th	2·6	(0·6)
	3·1			3·0	
	3·4	*3·1*		3·2	*2·9*
12th	3·0	(0·3)	27th	3·1	(0·2)
	3·3			3·0	
	3·1	*3·1*		3·2	*3·1*
13th	3·0	(0·2)	28th	3·1	(0·2)
	2·8			3·0	
	2·8	*2·9*		3·2	*3·1*
14th	3·0	(0·4)	31st	3·1	(0·2)
	2·8			3·2	
	3·2	*3·0*		3·3	*3·2*
17th	3·0	(0·9)			
	3·9				
	3·2	*3·4*			

TABLE 24 (*contd.*)

June	Samples	Mean/Range	June	Samples	Mean/Range
2nd	3·0 3·1 3·1	(0·1) *3·1*	21st	3·0 3·0 3·2	(0·2) *3·1*
3rd	3·3 3·2 3·3	(0·1) *3·3*	22nd	3·1 3·2 3·0	(0·2) *3·1*
4th	3·0 3·2 3·1	(0·2) *3·1*	23rd	2·6 3·3 4·0	(1·4) *3·3*
9th	2·7 2·7 3·0	(0·3) *2·8*	24th	2·7 3·0 3·4	(0·7) *3·0*
10th	3·3 3·3 3·5	(0·2) *3·4*	25th	3·1 3·5 3·2	(0·4) *3·3*
11th	2·9 3·1 3·0	(0·2) *3·0*	28th	3·0 2·4 3·3	(0·9) *2·9*
14th	2·3 3·2 3·2	(0·9) *2·9*	29th	2·7 3·2 3·0	(0·5) *3·0*
16th	2·5 3·4 4·0	(1·5) *3·3*	30th	3·0 3·4 3·6	(0·6) *3·3*
17th	3·2 3·2 3·6	(0·4) *3·3*			
18th	3·1 3·6 3·3	(0·5) *3·3*			

$\bar{x} = 3{\cdot}0$. $\bar{w} = 0{\cdot}49$.

(1) A group of values (n) per day, for example, 3 in Table 25, is first decided upon. That is, samples are taken from the sand mill three times a day and the values tabulated.

TABLE 25. CONTROL CHART LIMITS FOR SAMPLE AVERAGE

Sample size	For inner limits	For outer limits
n	$A'_{0\cdot025}$	$A'_{0\cdot001}$
2	1·23	1·94
3	0·67	1·05
4	0·48	0·75
5	0·38	0·59
6	0·32	0·50

(2) The next procedure is to obtain the arithmetic mean $\bar{x}$ for the three samples per day.

(3) The range w is also tabulated which is the difference between the maximum and minimum value of the day.

(4) The grand mean $\bar{\bar{x}}$ is the arithmetic mean of all the $\bar{x}$ values over the period of time which has been chosen as the satisfactory period of observation. The period chosen must be such that scrap casting is at a minimum, for example, 3% due to the effect of mould material.

(5) The mean range value $\bar{w}$ is the arithmetic mean of all the $\bar{w}$ values over the period.

The control charts are then plotted by using the property value as the ordinate and the period as the abscissa (Figs. 46 and 47). First the grand mean $\bar{\bar{x}}$ is noted but a line is not drawn. Then lines are drawn to show the control limits. There are two control limits, viz. the outer and the inner control limits on either side of the grand mean value, the outer control limit being the furthest away from the grand mean value. One method of calculating these limits is by using Table 25. Thus:

$$\text{Inner control limit} = \bar{\bar{x}} \pm (A'_{0\cdot025} \times \bar{w}). \qquad (21)$$

$$\text{Outer control limit} = \bar{\bar{x}} \pm (A'_{0\cdot001} \times \bar{w}). \qquad (22)$$

For the charts drawn in Figs. 46 and 47, the number of observations n was 3. Therefore, from Table 25:

$$A'_{0\cdot025} = 0\cdot67.$$
$$A'_{0\cdot001} = 1\cdot05.$$

The control limits for moisture (Fig. 47) is obtained as follows knowing $\bar{x} = 3\cdot0$ and $\bar{w} = 0\cdot49$ (Table 24).

Inner control limit:

$$\begin{aligned} &= \bar{x} \pm (A'_{0\cdot025} \times \bar{w}), \\ &= 3 \pm (0\cdot67 \times 0\cdot49), \\ &= 3 \pm 0\cdot33. \end{aligned} \tag{23}$$

Outer control limit:

$$\begin{aligned} &= \bar{x} \pm (A'_{0\cdot001} \times \bar{w}), \\ &= 3 \pm (1\cdot05 \times 0\cdot49), \\ &= 3 \pm 0\cdot50. \end{aligned} \tag{24}$$

The inner control limits are shown by drawing lines at 3·33% and 2·67% moisture. The outer control limits are drawn at 3·5% and 2·5% moisture.

The grand mean $\bar{x}$ for the green compression values is 5·6 and $\bar{w}$ has a value of 0·85 (Table 23). The limit values are obtained as follows:

Inner control limit:

$$\begin{aligned} &= \bar{x} \pm (A'_{0\cdot025} \times \bar{w}), \\ &= 5\cdot6 \pm (0\cdot67 \times 0\cdot85), \\ &= 5\cdot6 \pm 0\cdot57. \end{aligned} \tag{25}$$

Outer control limit:

$$\begin{aligned} &= \bar{x} \pm (A'_{0\cdot001} \times \bar{w}), \\ &= 5\cdot6 \pm (1\cdot05 \times 0\cdot85), \\ &= 5\cdot6 \pm 0\cdot89. \end{aligned} \tag{26}$$

The control limits are drawn by using eqns. 25 and 26 in a similar manner as those obtained for the moisture values. For control purposes the arithmetic mean $\bar{x}$ for each day is plotted; the individual values are also shown. All values should be within the inner control limits. Some isolated cases may be overlooked whose values lie in the band defined by the outer and inner control limits. If the values cross the outer control limits, all the probable variables in the sand plant system must be investigated.

Control charts must, however, be interpreted according to the performance of the moulding material. For example, for the moisture chart (Fig. 47) water content of the sand above 3% has been found to clog the sand hoppers and to give rise to certain casting defects and, consequently, from July onwards the moisture content of the sand is kept between 3 and 2·5%. That is, only the control limits below the $\bar{x}$ line are used from July onwards. The green compression values are kept higher than the corresponding $\bar{x}$ value in Fig. 46. In fact, the effect of decreasing the moisture has been to produce sands with values consistently between the grand mean and the upper outer control limit value in Fig. 46. It will be necessary to obtain fresh values of limits based on the observations between July and December.

Mouldability Controller

A good free-flowing sand with a minimum amount of moisture ensures a sand which rams easily and is found to produce good castings. The amount of moisture to give a constant green bond will depend on the presence of other elements in the sand and control charts for moisture and green strength cannot prevent the use of the mix or mixes for moulding and core making if the properties are not suitable. This is so because, of necessity, measurements of properties take time and the mould material by then has been delivered to the moulding hoppers and a degree of control on the sand before it is discharged onto the conveyors is certainly desirable. A development in this direction is the mouldability controller.

The mouldability controller is based on the principle that the mouldability of a mould material is determined by the toughness of the sand in the unrammed state which is measured by determining the ability of the product from the mill to bridge a slot in a vibratory feeder trough. The mouldability controller is attached to the casing of the mixing pan of the sand mill and comprises a slotted vibratory feeder trough and a system of photocells (Fig. 48). As the sand is mixed in the mill a sample is withdrawn

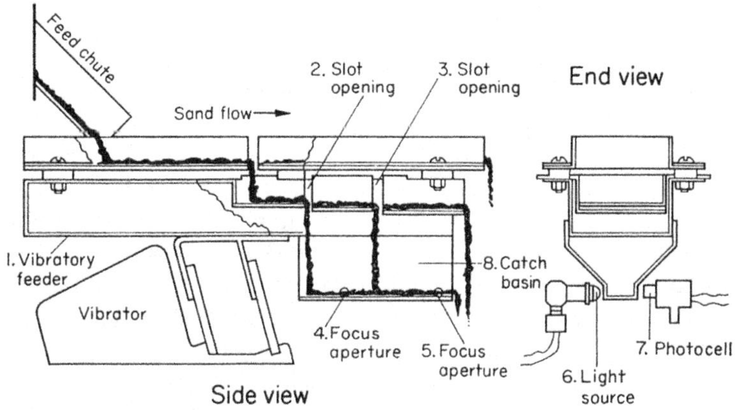

(*Courtesy Ridsdale and Company Ltd.*)

FIG. 48. Mouldability controller.

through a hole in the casing and falls onto the vibratory feeder 1. The sand then moves forward to the slot openings 2 and 3. When the sand is dry it falls through the slots and blocks off the focusing apertures 4 and 5. Two light sources 6 are directed towards two photocells 7, and the dry sand in the catch basin 8 interrupts the light beams. Through a system of relay and amplifier, two water valves controlling the water supply to the mill are opened. As the sand becomes wet and at an optimum mouldability, the slots 2 and 3 are closed and apertures 4 and 5 are clear to operate the photocells which close the water valves. The slots 2 and 3 are so adjusted that the water valves will shut off in sequence. The focus aperture 4 controls a rapid water addition while opening 5 adjusts the water to the final fine limits.

CHAPTER 17

RECORDS

MOULD and core materials play an important role in producing such serious casting defects as mould erosion, metal penetration, scab, hot tear, etc. Although, as discussed in previous chapters, certain correlations have been put forward between the properties of steel foundry mould and core materials and casting defects based on investigational laboratory work under controlled conditions, there is a paucity of data from actual shop floor experience. One difficulty in collating information from foundries is, probably, the lack of time because the castings must be produced and dispatched according to very tight schedules. Many foundries have made attempts to mark castings and note the properties of the mould and core materials from which they were manufactured. However, a careful examination of the castings subsequently has not often been possible because of a lack of personnel to follow the components before they were fettled and repaired.

It is certainly necessary to establish norms for acceptance tests for mould and core materials, and in conjunction with planned laboratory work valuable empirical data from the shop floor should be obtained to find correlation of properties and casting defects. A possible way is to have suitable inspecting staff who would work in close liaison with the sand controller and records should be kept along the lines suggested below.

The various sands should be numbered as shown in Table 26. The green, dry and high temperature properties should then be tabulated regularly as shown in Tables 27 and 28. It is then necessary to keep a record of casting defects along similar lines as shown in Table 29, giving as much details of weights, type of steel, etc., as

TABLE 26. PRODUCTION SANDS

Mix No.	Composition
1 (dry sand)	Natural sand A: 100%
2 (dry sand)	Natural sand B: 100%
3 (dry sand)	Natural sand A: 33%; natural sand B: 33%; silica sand: 34%
4 (dry sand)	As mix 3, but with 60 lb fireclay per 1200 lb mix
5 (green sand)	Wyoming bentonite: 4%; starch: 0·75%; remainder silica sand
6 (green sand)	As mix 5, but with 15 lb molasses per 1500 lb mix
7 (dry sand)	Natural sand A: 350 lb; natural sand B: 350 lb; silica sand 550 lb; glyso 37 lb; fireclay 50 lb.
8 (dry sand non-siliceous)	Chromite 112 lb; zircon flour 10 lb; ball clay 4 lb; Wyoming bentonite 4 lb; core oil 2 pints.

TABLE 27. GREEN AND DRY PROPERTIES OF PRODUCTION SANDS

Mix No.[1]	Green compressive strength (lb/in²)	% Moisture	Green perm. No.	Shatter index (%)	Dry compressive strength (lb/in²)
1	13·7	8·0	27	92	220
2	10·9	9·9	28	94	146
3	10·7	9·2	27	99	142
4	8·9	8·9	28	99	156
5	6·8	3·3	170	90	173
6	6·8	3·5	180	85	64
7	6·4	8·7	28	93	270
8	4·7	2·8	22	74	–

N.B. (a) moisture at 120°C. (b) dry compressive strength at 120°C for 2 hr.

[1] Composition of mixes same as in Table 26.

possible. The actual castings, moulding methods, gating system, etc., will be obtained if a column is used to mark the methods card number. Some general conclusions as follows should then be drawn.

TABLE 28. HIGH TEMPERATURE PROPERTIES OF PRODUCTION SANDS

Mix No.[1]	Hot compressive strength (lb/in^2) at soaking time, minutes (furnace temperature 1200°C)			Refractoriness % linear change at 1550°C	Thermal shock at 1350°C
	3	4	10		
1	590 (sh)[2]	910 (sh)	>1200	−1·72	Slight cracking after 2 min.
2	255 (sh)	445 (sh)	1050 (sub)	+1·50	Slight cracking after 1 min. One large crack after 2 min.
3	270 (sh)	245 (sh)	1020 (sh/sub)	+2·14	Began cracking after 70 sec, but only slightly cracked after 2 min.
4	180 (sh)	395 (sh)	1000 (sh/sub)	+2·98	One large crack after 2 min.
5	45 (sub)	30 (sub)	65 (sub)	+7·50	No cracks after 3 min.
6	—	—	—	+6·63	Severe cracking after 2 min.
7	90 (sh)	110 (sh)	80 (sh)	+3·39	Slight cracking after 2 min.
8	180 (sh)	320 (sh)	905 (sh)	−2·73	—

[1] Composition of mixes same as in Table 26.
[2] sh = shear type failure, sub = failure by subsidence. sh/sub = mixed appearance, while failed.

TABLE 29. RECORDS OF CASTING DEFECTS

Mix No.[1]	Casting	Steel specification	Approximate weight (cwt.)	Method	Remarks
3	Half liner	Medium carbon	2·5	Core	Burnt-on sand and mould erosion near runner
3	Blank	Medium carbon	7·0	Core	Small scab
3	Toothed pinion	Medium carbon	22·7	Core	Slight burnt-on sand in teeth
4	Toothed pinion	Medium carbon	34·5	Strickle, zircon facing	Clean casting
3	Mantle	Manganese	7·5	—	Slight metal penetration
5	Plate	Manganese	0·5	No core	Erosion near runner
5	Cover	Low carbon	—	—	Erosion
4	Wheel	Low carbon	27·0	Deep cod	Scab; slight burnt-on sand
4	Barrel	Low carbon	27·0	Large oil-sand core	Burnt-on sand
3	Roller	Low carbon	1·1	Deep cod	Scab; burnt-on sand
–	Roller	Low carbon	1·1	Chromite facing	Slight lapping effect

[1] Composition of mixes same as in Table 26.

Mould Erosion. This is prevalent while using manganese steel plate type castings, the defect being near the ingate. The properties of the green mould material as used show optimum green and dry compressive strengths, high shatter index and green permeability; positive linear expansion and good resistance to thermal shock; low hot strengths.

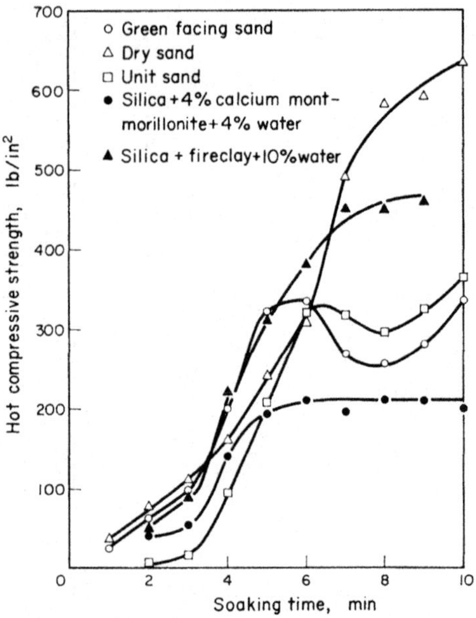

FIG. 49. Hot strengths of production sands at 1050°C.

Burnt-on Sand. Where this defect is prevalent the sand properties show high green strength; low permeability; high shatter index; reasonable dry strength; high hot strength; positive linear change; some lack of resistance to thermal shock.

It is noted that a high permeability mix causes mould erosion confirming the conclusions of investigational work. Mould erosion also results when the material possesses low hot strength suggesting that loss of bond strength at high temperatures may cause mould erosion, especially at points of persistent flow such as at or near the ingates.

Some of the properties can, perhaps, be correlated with the phenomenon of burnt-on sand. For example, high green strength

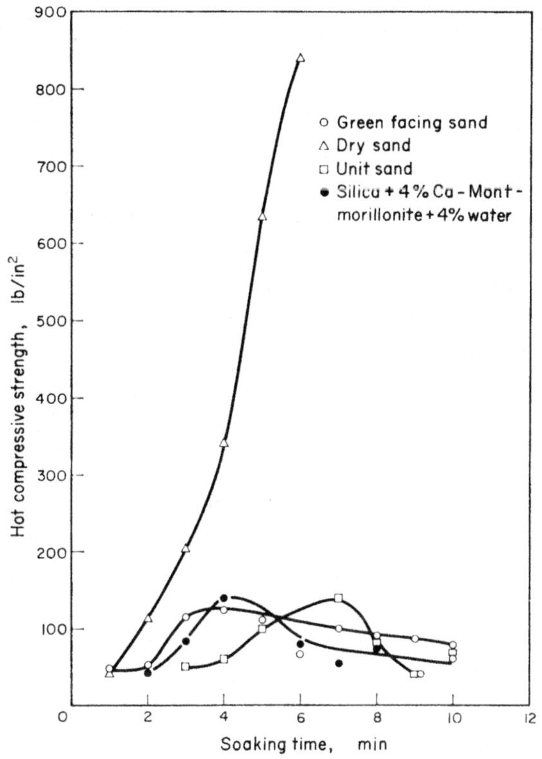

FIG. 50. Hot strengths of production sands at 1200°C.

together with high shatter index should decrease the flowability of the moulding mix. This would result in pockets of loosely rammed sand resulting in metal penetration under ferrostatic pressure, but this has not been experienced in this series of observations. Possibly a high hot strength and failure by shear causes sand particles to give the effect of sand rain when the mould is under high stresses. These detached sand grains may deposit on the hot skin of the casting giving the defect of burnt-on sand.

Weekly reports such as are shown here should be kept over a prolonged period of time and, possibly, a correlation may eventually result between casting defects and the properties of mould and core materials.

The hot strengths of production sands should be measured regularly. A few examples are shown in Figs. 49 and 50. An interesting feature of these graphs is that the green sands show a dip in strength around 800°C. This is probably due to the presence of starch which volatilizes, and is seen to be absent in a green sand bonded with a 4% calcium montmorillonite only.

It remains to be established whether it is the hot strength peak or the rate of strength development that will be the criterion in correlating hot strengths with casting defects. Experience of the present author with production castings so far has shown that the dry sands are more conducive to hot-tear defect than the green sands. It is noted (Figs. 49 and 50) that the dry sand has a steeply rising hot strength at both temperatures, whereas the slopes of the strength curves are shallower with the green sands. However, the peak strengths are also higher with the dry sand than the green sands.

INDEX

Printed in Dunstable, United Kingdom

91221476R00097